1
23
456
789

D0524828

Winning With Statistics: A Painless First Look at Numbers, Ratios, Percentages, Means, and Inference

Richard P. Runyon
C. W. Post College of
Long Island University

Addison-Wesley Publishing Company
Reading, Massachusetts · Menlo Park, California
London · Amsterdam · Don Mills, Ontario · Sydney

Cartoons by Gary Fujiwara.

 Printed in the United States of America. Published simultaneously in Canada. Library of Congress Catalog Card No. 76-17720.

ISBN 0-201-06654-8
BCDEFGHIJ-AL-7987

To Mom and Dad who have taught me that there is some element of humor in virtually all life situations.

Preface

Caveat emptor! This is not the usual type of statistics book. It is written to be plainly and unabashedly entertaining. But don't get me wrong. I am not denying that it is also intended to teach. What I am saying is that I refuse to subscribe to the view that statistics is a deadly serious subject that must be taught in a deadly serious, sober, and somber manner. I have never taught a course in that way and I don't see why textbooks must be a series of exercises in practiced tedium.

Of course, not everyone agrees with me. It would be one hell of a crazy world if they did. But let me defend my position. Any of you who has taught a course in Statistics has surely sensed the anxiety that permeates the classroom during the first few class sessions. The sense of foreboding is not in the least lessened by the student's first glimpse of the textbook, replete with strange and incomprehensible hieroglyphics and ponderous verbiage. In fact, in my experience, most withdrawal (psychological as well as physical) occurs during this first week. This is not at all surprising.

Over the years I have developed a strategy to counter this acute statistical trauma. I devote the first week or two of every undergraduate statistics course to statistico-therapy. What is statistico-therapy? It is this book — a mini-course in which I introduce many of the key concepts of statistics. It is the way in which these ideas are introduced that gives statistico-therapy its unique characteristics. I try my hardest to entertain the students, to show them that statistics can be fun and *is* relevant, and to teach them to relax in the presence of frightening stimuli.

Through the years, my students have responded favorably. Even today, when some come back to pay me a visit, they frequently recite back to me some of the stories I told in class and nostalgically recall some of the humor. However, it never occurred to me to put this minicourse down on paper.

The idea, then, is to relax the students; provide a broad overview of what statistics is about; develop an appreciation of the sweeping scope of statistical inquiry; and provide vivid, amusing, and easily remembered examples that will be accessible to recall throughout the course.

Let me wind this up with one story that illustrates my teaching philosophy. As many of you know, I have coauthored numerous books on statistics during my professional career. I still remember painfully the difficulty I had in finding a publisher for the first one. Admittedly, the first book (*Fundamentals of Behavioral Statistics*) was somewhat different. It included a smattering of humor. One publisher received a favorable review from one of its readers. However, the reader inserted one paragraph that I found unacceptable: "The most serious objection I have to this manuscript is the author's gross and elephantine sense of humor. Statistics is a serious subject that must be treated in an objective and detached fashion. Some of the things he says are, admittedly, humorous. These are the sort of things that can be presented in the classroom but should not appear in a book." I rejected his philosophy and withdrew the manuscript from consideration. The present manuscript is further testimony to my rejection of his pedagogical philosophy.

I wish to thank Nancie Brownley for laughing occasionally as she typed the manuscript and on other occasions saying, "I don't like the way that reads. I think it would sound better if . . . ."

November 1976 R.P.R.

Contents

One — The Two Faces of Statistics (Page 1)

Two — Don't Talk to Me About Numbers; People Are What Count (Page 11)

Three — The Disembodied Statistic (Page 25)

Four — Bias, Bias, Everywhere (Page 37)

Five • Gaffing with the Graphic (Page 55)

Six • Ratios, Proportions, Percentages, Index Numbers, and Hocus Pocus (Page 73)

Seven What Does the Mean Mean? (Page 93)
Eight A Standard Deviation is Not a Sexual Perversion (Page 105)
Nine We've Been Going Together for a Long Time but You Still Don't Turn Me On (Page 129)
Ten Having Fun with Statistics: The Probability Game (Page 157)
Eleven Make Me an Inference (Page 177)
Twelve You Can Prove Nothing Safe (Page 189)

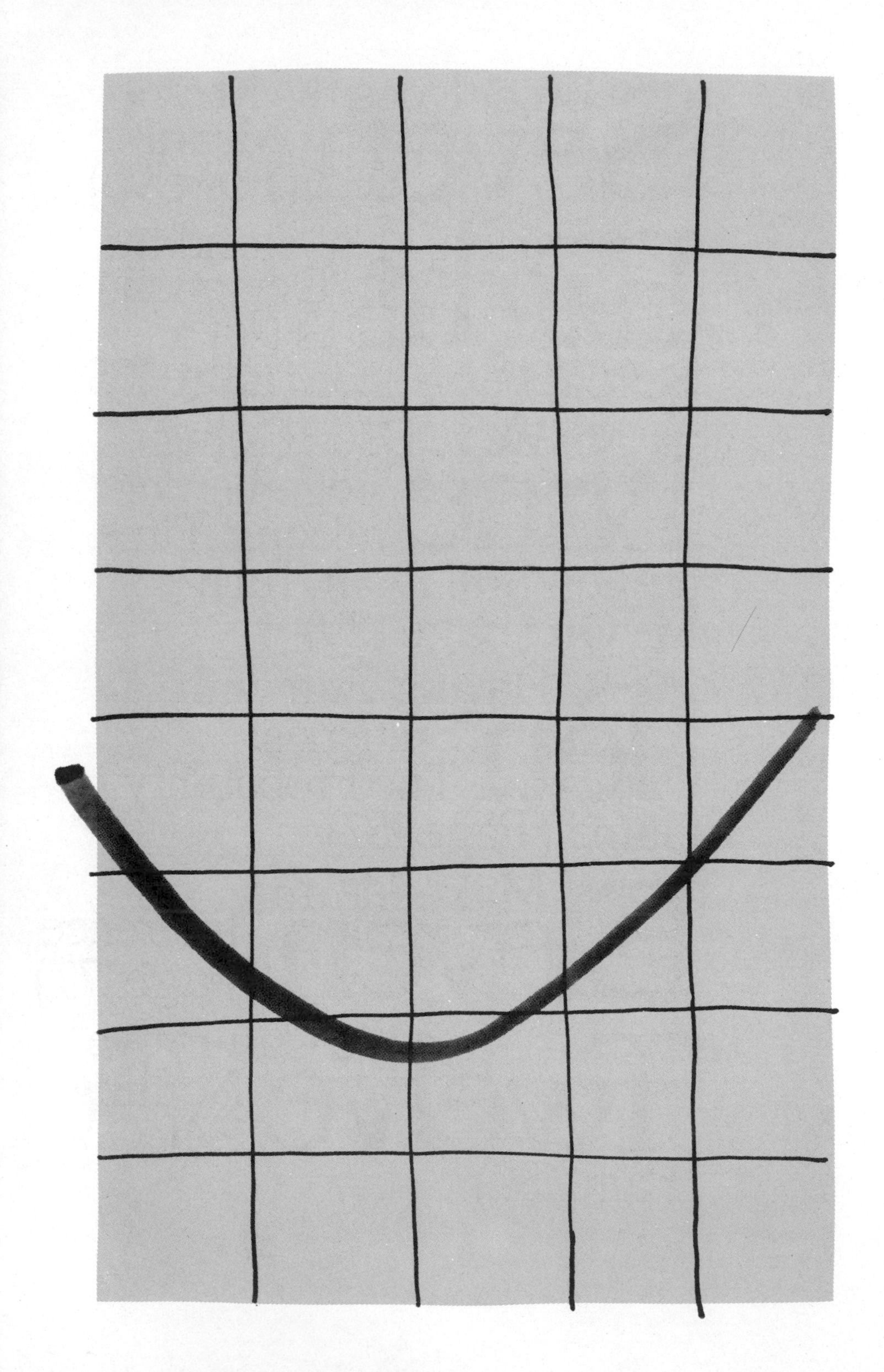

Chapter One: The Two Faces of Statistics

What would you think of someone who was conceived and born in a gambling casino in France, obtained an allowance during childhood by associating with villains and cutthroats, experimented with alcoholic beverages before attaining puberty, showed a deep fascination with manure during the adolescent years, and now in early adulthood, is fooling around with poisons and keeping company with some of the biggest liars around town?

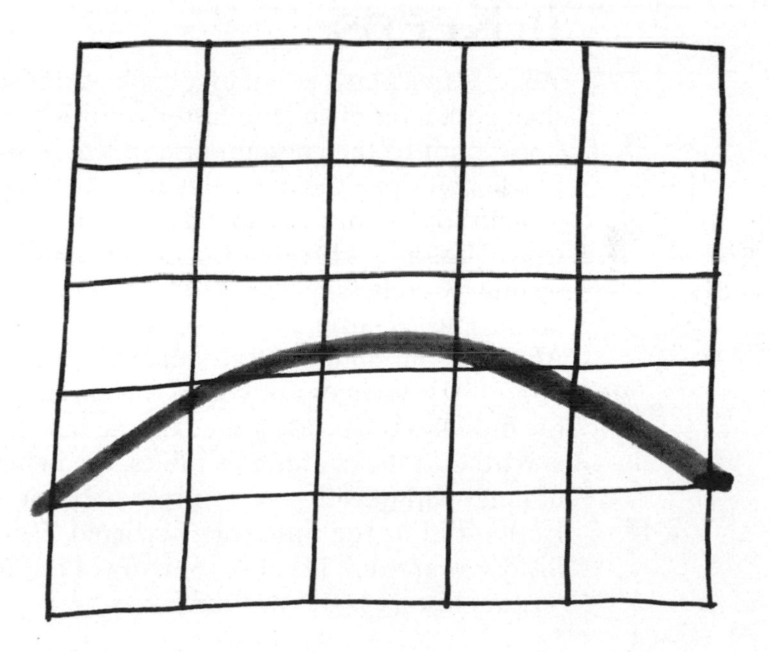

Who is this seedy character? That new grownup kid on the block? No. The Godfather? Public Enemy No. 1, 2, or 3? No to all of these. The unsavory character whose biography I just encapsulated is not made of flesh, blood, muscles, sinews, and bone. Nor is he usually given credit for having much of a personality. In fact, most people would describe his personality in terms that are vastly different from the ones I would use. According to popular notions, he is cold, indifferent, unfeeling, devious, and two-faced. Of course, now you know the character of whom I have been speaking. Statistics. Who else?

Statistics Is Known by the Company It Keeps

What better place to begin a book of this sort than with a bit of the life history of this misunderstood giant of the contemporary scene? Now please don't put the book down, either literally or figuratively, just because I mention that nasty word "history." History need not be a dull and seemingly endless recitation of dreary dates and events in the continuing struggle of our species. After all, the human affairs that history books relate—the intrigues of court, the clashes of the muscle and spirits of men on the battlefields and around the conference tables, the beheading of kings and queens—were pretty vibrant and frothy stuff at the time they occurred. I guess history seems dull because so many of the people writing history books are dull.

So it is with statistics. Dull not because of its content nor because of the situations to which it applies. Dull because it is typically presented in a dull manner. But statistics has always led a double life. Indeed, the fellows with whom it has shared its bed would have fascinated the master of the risqué, Guy de Maupassant. Let's see why.

One of the first thrusts of statistics into the affairs of humanity came at the gaming tables of France during the middle of the seventeenth century. Two French mathematicians, Pierre de Fermat and Blaise Pascal, exchanged a series of notes concerning games of chance and wagering. The odds and probabilities that they calculated are still used in such farflung places as Las Vegas, Monte Carlo, and Puerto Rico. In fact, it is because the odds in various games of chance have been worked out with such consummate precision that, as we shall see in Chapter 10, gambling casinos are among the most scrupulously honest business activities on the contemporary scene. This next statement may seem like a bit of hyperbole but let me say it. I believe they are even more honest than oil companies or the Watergate conspirators.

Be that as it may, let us take a brief look at the circumstances that prompted two respected mathematicians to dabble in the dubious delights of dancing dominoes, otherwise known as dice. It seems that a very wealth French playboy, known as Chevalier de Méré (which, incidentally, comes frighteningly close to meaning "horse manure"), was adding considerably to the family coffers by cajoling suckers into accepting the following bet: "Even odds that if I roll a die four times, I will obtain at least one six." He reasoned as follows: "If I roll a die once, the chances of it turning up a six is one in six. If I roll it twice, the chances will double—two in six. At three tosses, it will triple, yielding odds of fifty-fifty (three in six). But at four tosses, I will clean up. The chances of obtaining at least one six is four in six. Four favoring, two against . . . that's two to one in my favor!" Actually, his reasoning was wrong as he would have noted if he had kept a careful record of his winnings. He would have found a ratio of winning to losing of about three to two instead of two to one. Or he could have carried his reasoning two steps further: "If I toss the die five times, the chances of getting at least one six is 5/6. If I toss it six times, my chances are six in six. That's absolute certainty!" Anyone the least bit familiar with the behavior of dancing dominoes knows that certainty is not one of the better known attributes of dice. For the record, the probability of obtaining at least one six (or any other number) is about two to one only when a die is tossed six times, rather than four times as de Méré reasoned.

So in spite of his faulty reasoning, de Méré was raking in money hand over fist. But his name might never have made the history books nor is it likely that Fermat and Pascal would have had the opportunity to get into the act if de Méré had not overreached himself in his greedy quest for ever more booty. He began a simple variation of his highly profitable game. He reasoned, "If I toss two dice, the probability that I will get a twelve (two sixes) is one in thirty-six. If I toss these dice twenty-four times, the likelihood of getting at least one twelve is 24/36 or 2/3. That's again two to one in my favor." But alas, poor de Méré began losing his shirt. Unbeknownst to him, the odds had now turned in the opposite direction—against the house and in favor of the bettor. Forlorn, bewildered, confused, and with all the cocksureness knocked out of him, de Méré turned pleadingly to the mathematician Blaise Pascal. "What went wrong?" he cried in anguish.

The rest is history. Pascal and Fermat solved de Méré's quandary and probability theory took a giant leap forward into the affairs of humanity.

Later in the seventeenth century, Halley published his famous Life Table, which was the first actuarial table and the prototype used by the thousands of insurance companies that have since gained respectability. It was not always thus. The Halley tables actually represented the emergence of insurance as a legitimate enterprise. Prior to that time, insurance on merchant ships was among the most widely speculative acts of the financial world. There were almost as many causes for losing ships as there were ships. They could founder in storms or be smashed asunder while navigating around the extremities of South America or Africa. They were the helpless prey to the navies of unfriendly nations or to buccaneers (often one and the same). They might vanish in the still mysterious "Devil's Triangle." And then there were the inevitable mutinies, fires at sea, encounters with the Loch Ness monster and Bigfoot. Anyone had to be out of his mind or flagrantly disreputable to insure a ship ("I don't care what Columbus discovered!

One of them is bound to fall off the edge of the earth one of these days. Mark my words!"). Insurance has sure come a long way since the days when it was no more respectable than the gaming tables, and considerably less respectable than the oldest profession where the going rates were more stable and more readily ascertained.

Ah, but not all greatness and miracles were reserved for the dim past. Toward the turn of the present century, a statistician by the name of William Gosset was hired by a prominent brewery in merrie olde England to set up and maintain quality control procedures in the brewing of ale. In addition to ale, many statistical ideas fermented in that factory and out of Gosset's labors came the description of a family of mathematical distributions that introduced a branch of inferential analysis known as "small-sample statistics." We won't concern ourselves at this time with the details of Gosset's statistical discoveries except to note in passing that quality control processes are not restricted to the brewing process. There is hardly a quality control procedure in use today that does not owe a debt of gratitude to Gosset's pioneering work.

Next came the "manure phase" of statistical development. Early in this century, the application of scientific methods to the field of agriculture had taken giant steps forward. The burgeoning population and the relative decline in arable land made it mandatory that the success and failure of crops be removed as much as possible from the vicissitudes of nature. Carefully controlled research was undertaken to ascertain the effects of innumerable variables or conditions on the yield of crops. How did horse manure compare in effectiveness with cow chips or goat pellets and how do these, in turn, stand up against several man-made fertilizers that were appearing on the market with increased frequency? To answer questions such as these required a new form of statistical analysis. A brilliant British statistician and agronomist, the late Sir Ronald A. Fisher, was equal to the challenge. He developed a technique known as the Analysis of Variance (ANOVA). It is now the most widely used form of statistical analysis in a broad spectrum of scientific disciplines.

More recently, rapid approximate statistical methods were developed by the late Frank Wilcoxon for American Cyanamid, where he served as research and control statistician. And we all know what American Cyanamid does. It makes fungicides, pesticides, herbicides, and all of those other pleasant potions intended to rid us of the plants and animals that appear on the homeowner's "enemy list."

But wait a minute. Let's think over that biography again. Yes, just as I suspected. In all of these examples, statistics is playing the part of the good guy. Statistics has forced the gambling casinos to be scrupulously honest; it has given a respectability to insurance and, in so doing, has created millions of jobs; it has made possible high levels of quality control (where practiced conscientiously) in spite of the fact that many items—ale and beer included—are manufactured in the millions of units; it has shown agriculture the way to improve crop yields by quantum amounts over the past seven decades; and it has provided the techniques by which control over insect-borne diseases has been achieved. Not bad for a kid brought up by such a dubious crowd.

But alas. Many do not judge statistics by its accomplishments but rather by the company it keeps. Even I must admit that statistics has fallen in with a pretty bad crowd in recent years, albeit through no fault of its own. Politicians and advertising agencies have not been loath to perpetrate statistical lies if they advance the cause of the great perpendicular pronoun, in the one case, and the agency's bookings in the other. We have all heard expressions similar to the following: "A politician uses statistics in the same way a drunk uses a lamp post—more for support than illumination." Or, "There are lies, damn lies, and statistics."

But to blame statistics for all the lies perpetrated in its name is a bum rap. Statistics don't lie; people do. When used by individuals dedicated to finding truth, it is like a scalpel in the hands of the finest neurosurgeon; when put in the hands of a person wishing to prove something —whether or not it is true—statistics can be more like providing the surgeon with a guillotine. It severs the connection between reason and reality. It is precisely because of the widespread misapplication of statistical analysis that much of this book is dedicated to exposing statistical fraud. But it would be an injustice to direct our attention exclusively to the seamy side of statistical life. Therefore, much of the book will also deal with those aspects of statistics that caused the nineteenth century seer, H. G. Wells, to prophesy, with characteristic accuracy, that the ability to think statistically would some day be as important for good citizenship as the ability to read or write.

7 8 9 ÷ C
4 5 6 X CE
1 2 3 − %
0 • + =

Chapter Two: Don't Talk to Me About Numbers; People Are What Count

John Augustus Redington Pass sat quietly and contemplatively in a room that was still thick with the memories of the exhausted cigarettes, profane epithets, and roiling emotions. His outward calm belied the turmoil that raged within. For six months he had plowed all of his energies, all of his finances, all of his physical and spiritual being into running for mayor of Metropolis. Now his closest friend and campaign manager was advising him to drop out of the race.

"You know I'm not a quitter, Frank," he said, staring absently at a wisp of smoke rising lazily from a "dead" cigarette butt. "I have never been a quitter and I don't see why I should quit now."

"It's not a matter of quitting, John. You put up a good fight, a damn good fight. But the numbers are against you."

"Numbers? You speak to me of numbers? It's people we're talking about Frank. It's people who have been screwed from here to Timbuktu by those bastards in office. How can you talk to me of numbers?"

Frank remained calm, unperturbed. "Dear friend, it's people who vote but it's numbers that decide. The numbers are all against you. The machine has won."

The mayoral candidate flinched, almost imperceptibly. He felt vague, diffuse movements in his abdominal region, punctuated by a sharp stab of pain. "I'll have to take another Tum," he thought, "when no one is looking." He raised himself from the chair, walked quietly to the window. "Look out there, Frank. There are more than three million people in this city. A million registered voters. And you're telling me to throw in the sponge because a poll of one thousand voters shows Glotz out in front?"

"A scientifically conducted poll, John."

He felt his lips curl momentarily into the rictus of rage. But just as rapidly it was transfigured into a wan smile. "What am I getting so upset about? And what makes you so cocksure that Glotz has it in his hip pocket?"

"He has the numbers."

"Will you please stop saying that, Frank? He got six hundred and fifty out of a thousand people to say they'd vote for him. That's all. Do you realize that one thousand people is only one-tenth of one percent of our total electorate? Are you telling me that we're going to let one in every thousand voters decide whether Metropolis gets another four-year screwing?"

"The polls don't lie, John. Those one thousand respondents were not pulled out of a trash heap, you know. They were carefully selected so as to be representative of the registered voters in the city. Both sexes, all religious groups, every minority represented according to their numbers."

"Don't feed me that pollster crap, Frank. You know and I know that they can be wrong. They have been wrong. Remember the Truman–Dewey election? All the pollsters picked Dewey. Even that commentator Kaltenbalm, Kaltenblum, or whatever his name was, picked Dewey the morning *after* the election."

"Ah, but the pollsters who kept surveying right up to the end of the election saw that it was going to be a cliff-hanger, which it was."

"But they still picked Dewey and they were wrong. Now look, Frank, I don't want to sound like a surly brat who refuses to admit it when he's licked. I just don't think I'm finished. There are 999,000 voters we haven't heard from yet. Let's give some of them a chance."

"You have heard from them, John. Through the poll."

"Just a thousand. Only a thousand, goddammit. How can a thousand tell you about a million?"

"Have you ever watched your wife bake a cake? She sticks one lousy toothpick in the cake and judges whether or not it's done. She doesn't have to make a pincushion out of the thing. When the park department tells you the ice is safe for skating, they don't have to remove all the ice on the lake to see how thick it is. We're forever looking at samples and judging the whole from what we find. Right now, for example, I'm sampling your behavior."

"Oh, and what do you find?"

"I notice you're standing as far away from me as possible, your arms are folded tight against your chest, and your breathing is irregular. You're angry with me."

"By God you're right. Maybe there's something to what you say."

"Is there any announcement you'd like me to make to the media? Any decision one way or the other?"

"No. Let me sleep on it." He turned back toward his desk and the piles of correspondence that awaited his attention. Frank sampled his actions and judged that the meeting had come to an end, an inconclusive end at that. As he left the room, he heard his friend of many years mumble, "A thousand lousy voters. . . ."

This glimpse into a significant moment of the life of John Pass reveals two more facets of the field of statistics. There are two broad functions of statistical analysis. One is called the descriptive function. It is concerned with the collection and subsequent manipulation of data so as to yield such quantitative summarizing statements as: "Of 1000 registered voters polled, 650 expressed a preference for Glotz." The summarizing statistic here is the percentage—sixty-five percent favored Glotz over Pass.

But it's rare that we are interested in the descriptive statistics as such. The preferences of one thousand out of a million voters would hardly arouse anyone's interest as such. But if this sample could be considered representative of the population of voters in Metropolis . . . that would be something else again. The second function of statistics is concerned with the process of making inferences from samples to populations. Not surprisingly, it is referred to as *inferential statistics.* Let's take a closer look.

What Statistics Is All About

If you're going for a swim, you might just as well jump right into the water and get over that initial discomfort all at once. If you're going to be a statistician, the same principle applies. So we're going to make you a statistical consultant to the Littlejohn Screw Corporation. In the event that you do not keep up with these things, Littlejohn is the company that won a contract for $82,635,433.08 from NASA to manufacture precision screws and bolts for orbiting space satellites. For years it has boasted, "We're the company that screwed up the space program." I am sure you'll remember some of its TV, radio, and newspaper advertising campaigns. One of the most successful was aimed at keeping people from handling its screws—essential because Littlejohn screws are precision instruments and excessive handling can destroy finely tuned tolerances. I am sure you'll remember this one. A serious-minded middle-aged man dressed in a white coat approaches a group of ladies in a hardware store and lugubriously intones, "Ladies, please don't squeeze the screws." Another less successful campaign was rapidly dropped so that you might not have seen it. It showed a picture of a man using Littlejohn screws as a pillow. The pitchman proclaimed joyously, "Thousands of tiny pillows built into each Littlejohn screw. Littlejohn is the only screw that lets you sleep."

Well, whether or not you remember the advertising is unimportant. What is important is that you are the quality control expert. You are responsible for making sure that your company releases no screw except a good screw. At this very moment, your company has a contract to produce millions of those little buggers a day. The contract calls for these screws to be 50 millimeters in diameter with an error of no more than ± one millimeter. If more than one percent of a day's output exceeds these tolerance limits, a severe financial penalty is imposed that could, if repeated often enough, bankrupt the company. So the stakes are high, Mr., Mrs., or Ms. Expert. What do you do?

Take your time. Think this one through. Don't rush on my account. I have all the time in the world.

What's that? First thing you'd do is figure out how to measure those little critters? Good start. The measurement problem is the first one you must solve. If your data base—your raw measurements—is not accurate, you might just as well scrap the whole effort. That's where public opinion polls and surveys encounter their greatest difficulties. You can select the sample using the most advanced methods of selection and use incredibly sophisticated high-speed computers to analyze the results, but if a number of your respondents chose not to answer honestly, your results are as useless as last night's theater tickets. Or if you ask an ambiguous or leading question . . .

Let me illustrate. Back in the 1930s, farmers along the Ohio River were being wiped out year after year by the spring floods. It was proposed that the government build a series of dams for the purpose of flood control. When the farmers-of-annual-grief were asked the question, "Are you in favor of the government building dams for flood control?" the response was overwhelmingly favorable. But when the question was slightly reworded, "Are you in favor of the socialistic practice of the government building dams for flood control?" the farmers voiced a resounding "Nay!" So it is that we often respond to words rather than issues, claims rather than facts.

OK. Let us assume that you have found a foolproof way to differentiate the bad screws from the good screws. What do you do next?

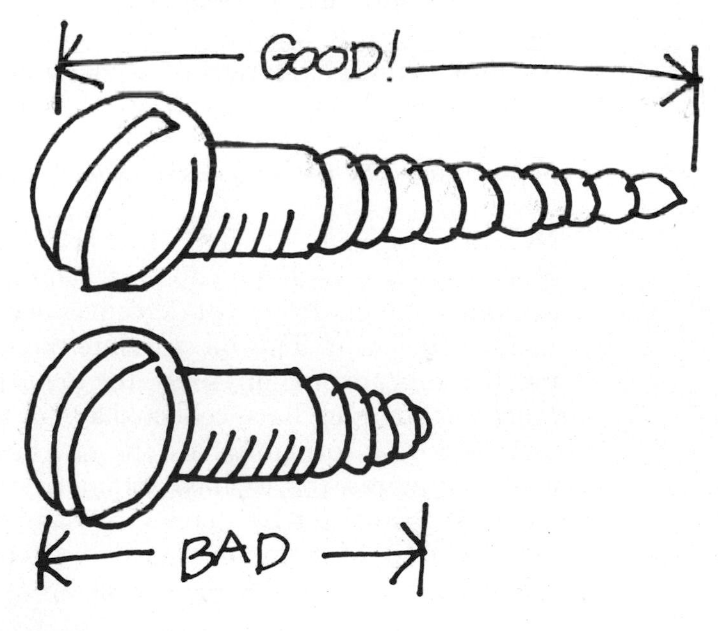

Find some way to select representative samples of the day's output? Good. Couldn't have stated it much better myself. But why not test every screw?

Now calm down. Please don't get so excited. I wasn't *telling* you to test the lot. I am only asking, "Why not?"

Good. You're absolutely right. There is no way that you could economically test the entire output of screws. The company would, as you say, go down the tube in no time at all. That's not exactly the way you phrased it? But I think I caught your meaning without offending anyone.

So what's your alternative? Random sampling procedures? Good. Excellent. And how would you do that? I understand. You are the boss statistician and you concern yourself only with broad strategies and avoid the everyday nitty-gritty details, right? So you would hire a statistical consultant? I'll buy that. Actually, I didn't want to go into random sampling at this point either. Suffice it to say that it is merely a method to ensure that each sample of a given number of screws is equally likely to be selected from what we experts call a *population*. In this case, the population is the entire day's output.

Now that you have collected a number of samples of screws, what would you do next? You would determine the proportion of defectives in each sample? Excellent. This takes us into the first function of statistical analysis—the *descriptive function*. After you have collected a mass of data, you want to manipulate the raw facts in ways that permit you to make summary statements. If you collect 100 screws in a sample and find that three of them are defective, you could summarize the results for the sample by saying that the proportion of defective screws is 0.03, or three percent. For other types of data, you may be interested in using other descriptive statistics, such as the mean, the median, or the range.

---

box 2.1 *Confusing a datum with a statistic*

How often do you hear the MC at a beauty pageant announce, "Her vital statistics are 36, 23, 34?" These aren't statistics. They're data. A statistic is a statement that summarizes a collection of measurements. It would be correct to say that the mean (a statistic often referred to as the average) vital statistics of all contestants are 36.4, 23.6, 34.3.

Similar confusion arises whenever a commercial makes such statements as "Come to Junk-heap TV Repair. Our service personnel have one hundred years of repair experience," or "In more than 1,000,000 grueling miles of testing, our automobiles had a repair record of less than one percent." What the announcers carefully left unsaid were the number of repairmen in the TV shop and the distribution of their work experience and the number of automobiles tested. For all we know, the boss may have had 25 years of experience, with each of 75 underlings able to claim only one. Also I would not be terribly impressed to learn that 100,000 new cars had been put through ten miles of testing each. How many breakdowns are likely to occur during the first ten miles in the life of a new car?

But your work doesn't stop there, does it? In many ways you're closer to the beginning than to the end. What must you do next?

Precisely. We want to estimate the proportion of defectives in the population (the whole day's output) of screws. If the sample proportions lead you to estimate a proportion of defectives greater than one percent, the company is in trouble. The engineers will have to go over the manufacturing process in the greatest detail to correct the trouble. On the other hand, if it appears safe to conclude that the proportion of defectives is less than one percent, we are in clover. We can continue to assist NASA in its enormous screw job. In any event, whenever we use the statistics taken from a single sample or a number of samples to estimate characteristics of the population, we are taking a giant inductive leap into the unknown from the known. This leap involves the second major function of statistical analysis—the *inferential* or *inductive function.*

In many ways inferential statistics is the most exciting aspect of the statistical ball game. Whenever you take an inductive leap, there is always an element of risk involved. Pollsters sometimes "call the wrong winner" as Clem McCarthy did some years ago when calling the Kentucky Derby. Occasionally a drug is released that is not really "safe" or "effective." The beauty of statistical inference is that, while admittedly involving the risk of an incorrect conclusion, it permits us to estimate precisely what the risk is. And we can raise and lower the risk, as the situation demands. But I'm getting ahead of myself. We'll be returning to this fascinating aspect of statistics in Chapter 11.

Those Mischievous Media Merchants

Confusion between the two functions of statistical analysis can be a great source of mischief. The TV pitchman solemnly intones, "Sixty percent of doctors interviewed prescribe WIB for the relief of tension headache." This "sixty percent" is a descriptive statistic. It could be based on as few as five interviews (if three say they prescribe WIB, that's sixty percent). But the trouble is that most people remember the statement that sixty percent of doctors—the whole population of doctors rather than the sample—prescribe WIB. They have made the inductive leap without realizing it. And the words of doctors carry much weight in our society . . .

Or maybe the statement is based on a larger and more representative sample. Let's look at this beautiful little bit of chicanery: "Of three thousand doctors interviewed, eighty percent prescribe WIB for the relief of tension headache." That's a lot of doctors. Most impressive! But let's look at the interview.

"Doctor, do you prescribe OAT for the relief of tension headaches?"

"For some patients, yes."

"COH?"

"Yes, for some patients."

"IRD?"

"On occasion."

"WIB?"

"Yes, on occasion."

The doctor prescribes WIB so it goes down as a "Yes" response. But nobody tells us that he or she may also prescribe everything else under the sun, including snake oil.

Or, how about this one: "Ninety-five percent of the doctors prescribe the pain reliever found in WIB." What is left unsaid is the fact that the same pain reliever appears in OAT, COH, and IRD. This is very similar to the claim, "Drivers taking the first three places at Indy wore Bliby's Bloomers." What we are not told is that every driver in the race wore Bliby's Bloomers. (They also gargled with STP.)

Then there is the other form of hocus-pocus that raises the non sequitur to the level of a fine art. Here's what you do. You cite research results that have nothing to do with the advertised use of the product but—pay attention, this is important—in so doing you must keep an absolutely straight face. Otherwise, the audience will discover the deception. My nominee for the all-time champion of the non sequitur is David Janssen who solemnly informs us that studies made by a leading university showed that Excedrin is more effective for "pain *other than headache*" (italics mine). He then concludes with cherubic innocence and sincerity, "So the next time you get a headache, try Excedrin." How many of you caught that non sequitur? Of course, if he had said, "So the next time you get cancer, try Excedrin," we would probably have noted the transgression.* Oh, if the participants weren't so damned serious and self-righteous about it, TV lies could be fun.

* Apparently others noted the non sequitur. After this section was written, a new commercial has appeared in which David Janssen says, "So the next time you feel pain. . . ."

"Is it true that they use only 10 percent of their brain during the course of a lifetime?"

Chapter Three: The Disembodied Statistic

Many of us old-timers remember the soap ad that proudly proclaimed "Ivory Soap (or was it Ivory Snow?) is 99.44 percent pure." I must admit that I was most impressed when I first heard it. Now, no longer an impressionable and callow youth, I am more impressed by the audacity of the claim than by the purity of the soap. This ad beautifully illustrates what I call the "disembodied statistic." The three essential ingredients of a disembodied statistic are: (1) it must be exceedingly precise; (2) it should provide no frame of reference that allows the reader to make comparisons; and (3) the definition of the main concept should be left ambiguous.

Let us face it. Nothing blows the mind more and is as unassailable as a precise statistic. 99.44 percent pure! That's hard to beat. But I must be honest with you. I have pulled a few classics in my time. On the average of 1.73562 times each year, a student will interrupt my lecture with a question such as: "Professor, is it true that we use only ten percent of our brain during the course of a lifetime?" Now I ask you, how do you answer a question like that? For years my approach has been to break down that statement systematically into its component parts. What do you mean by "use"? How does a person use his or her brain? Unless we can come up with some satisfactory definition of this word, the rest of the statement is sheer gobbledegook. Even assuming that we can arrive at some mutually satisfactory definition, we're not much better off. How do we measure the proportion of the brain that gets "used" in a lifetime? Could we stick thousands of electrodes on various parts of the brain to determine whether the underlying nerve cells are firing? Ah, but it is a neurophysiological fact that they will all be firing if they are not dead, that is, 99.650374109 percent of the time. Hey, we could go on with this analysis *ad nauseam*. I think the point is clear. The question is unanswerable and any attempt to provide a statistic is fraudulent, no matter how well intended.

How do I field this question nowadays? Why, I simply reply, "Your statement is absolutely false. The truth is that we use our brain only 8.45603 ± 0.000005 percent of the time." After giving the class a few moments to recover its composure, I continue: "I think that the question is really concerned with the efficiency with which we use our time and talents. The question of human efficiency is subject to some determination, within broad limits and under certain agreed-upon definitions of inefficiency." I then discuss various ways of measuring human performance.

A Few Examples

We are assailed by disembodied statistics with such frequency each and every day of our lives that I fear they have destroyed our critical faculties. We hear them, feel a bit uncomfortable about the meaning or accuracy of the claim, take a sip of beer (made with pure reprocessed sewer water), and permit ourselves to be mesmerized by the dancing images on the boob tube.

— The star athlete, recovering from a disabling injury, proclaims in genuine seriousness, "I am only about 40 percent right now. By next week I expect to be 60 percent, but my doctor tells me it will be another month before I am 100 percent." The TV commentator (not usually the most perceptive person in the world) is delighted. "You heard what he said. He's only 40 percent but 100 percent of his opponents quake at the mere sight of him. Just wait until he's back to 100 percent. Oh, boy!"

— The shill on the "Wee Wee Hours Movie Greats" proclaims, "Dodocs are more effective." And I cry out from the depths of my insomnia, "Than what? What do you mean by effective?" But he never answers and my protests are lost in the ether.

— I just went through a magazine and found advertising for eleven different brands of cigarettes. They all report the number of milligrams of tar and nicotine per "average" cigarette. But what does this tell you about the risk to your health? Moreover, assuming that lower quantities of tars and nicotine are less dangerous than higher quantities, the reader of the magazine would have to compare all the ads (as I have done in Table 3.1) before deciding on his or her

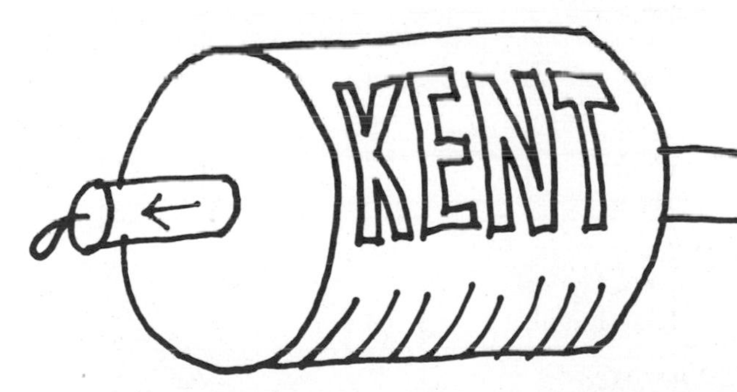

own brand of poison. Incidentally, I have heard a rumor that a new filter has been developed that completely blocks all tar and nicotine. It also blocks smoke. The only trouble is that 11.67 percent of the people trying to draw on the cigarettes become hernia victims.

The claims of advertisers, shills, and politicians remind me of a story I heard years ago in a Sunday sermon. It seems that there is some fabled restaurant district of some large American city (also fabled, I presume). One restaurant pro-

table 3.1 Amounts of tars and nicotine reported in separate ads in a single magazine. This list was compiled by me for your benefit. "Now that I have these statistics, what should I do with them?" you ask. I won't touch that question.

Brand	Mg/average cigarette Tar	Nicotine
Eve		
Filter	18	1.3
Menthol	18	1.3
Kent	18	1.2
Kool	17	1.2
Newport		
Kings	18	1.2
100's	20	1.5
Raleigh		
Extra Mild	14	0.9
Filter Kings	15	1.0
Saratoga		
120's	16	1.1
100's	16	1.1
Tarryton		
Kings	20	1.3
100's	19	1.3
True		
100's Menthol	12	0.7
King Regular	11	0.7
Viceroy	17	1.1
Virginia Slims		
Regular	17	1.0
Menthol	17	1.0
Winston	14	1.0

claimed in large letters, "THE BEST RESTAURANT IN THE CITY." The next on the block would not be outdone; "THE BEST IN THE STATE," it boasted. A third, larger and more garish than the other two, boldly stated, "WE ARE THE BEST RESTAURANT IN THE UNITED STATES." A fourth laid claim to the world's title. At the very fringe of the food-for-fun-or-ptomaine district stood a tiny, unimpressive hole-in-the-wall. It displayed a diminutive sign in its window:

the best restaurant
on this block

But I think the real culprit in the growth of the disembodied statistic is not the advertising agency, the unscrupulous businessman, or the mass media. The fault is in us. All too often we demand precise statements and we will not settle for less. There is something comforting and reassuring about a number, no matter how questionable the assumptions through which it was derived. It's comforting to the person perpetrating the number (it makes you look so sagacious) as well as to the person who is happy to add your magic number to his or her arsenal of disembodied statistics (sometimes called trivia by the unappreciative).

Let me illustrate, again from my own life. Quite a few years ago, a close friend and chemist (Dr. Lawrence Rocks) and I wrote a book called *The Energy Crisis* (Crown Publishers, 1972). In it we engaged in a number of broad statistical exercises with energy data and made some cautious (our word) extrapolations into the future. Early reviewers referred to these extrapolations as "alarmist." As one prediction after another was confirmed in the real world, later reviewers referred to these same projections as "overly conservative." We also made a number of somewhat less cautious political extrapolations based on our understanding of the facts of international politics. (For example, we stated that the Arab nations would use the conflict with Israel as a pretext for declaring an embargo on oil. The result would be exorbitant costs for petroleum products, inflation, and unemployment, particularly in the automotive industry.) At first, this book went over like a buffalo chip in a punch bowl.

But then things began to happen in the Middle East. Suddenly, but for a short period of time, Larry and I became genuine 62.53-percent celebrities. We received calls from TV and radio stations almost daily, asking us to submit to interviews. Except for a few shows, the TV circus was unreal: "Dr. Runyon, tell us what the energy crisis is all about (you have one minute)" or "Dr. Rocks, what is the future of alternative energy sources? (Please keep it under 30 seconds.)" The radio shows were more relaxed and permitted us to answer in considerable depth. In fact, I was on one show for five hours nonstop. But throughout all the interviewing, one demand persisted, either directly or by implication: Give us precise statements! We like 99.44 percent. It sounds knowledgeable. Finally, on one radio show, I couldn't take it any more and gave a very precise statistic. The following is a general account of one segment of the interview. (It is not exact since I do not have a tape recording of that show now. To tell you the truth, I can't even remember the name of the show.)

"Professor Runyon, tell us what percent of our total energy we can expect to derive from the sun by the year 1990."

"Well, that is a difficult question. The answer depends on so many factors—how much money the government allocates to solar research, the quality of research, the economics of providing solar energy . . ."

"Excuse us for interrupting, professor, but let's cut through the hedges. I would like to know, and I am sure our listeners would also like to know, just how much solar energy we can expect to have by 1990."

"Well, that is what I was getting to."

"I understand but let's cut through all the ifs, ands, and buts. How much can solar energy contribute to our nation by 1990?"

"Under the right conditions, we could have 99.44 percent of our total energy from the sun." (I actually used the Ivory soap figure; I really did!).

"Why, that's almost 100 percent. That's the highest estimate I have heard yet. You really believe this is possible?"

"Absolutely, given the right conditions."

"Amazing. What are the right conditions?"

"We need a genuine biblical miracle."

To any among you who may have been one of the 27.463 people listening to that interview, I want to apologize here and now for my apparent flippancy. I am not as proud of that answer as the one I gave to Jerry Stiller (of the Stiller and Meara team) on the *Mike Douglas Show*. Stiller had just sneezed on camera. He quickly asked, "Is there any way of harnessing that?" I promptly replied, "We could attach a windmill to your nose."

Windmills on noses or not, the energy crisis—like so many crises that have surfaced over the past decade—has already built up an impressive repertoire of disembodied statistics.

Let's look at one example. While I was participating in an energy conference at the Watergate complex in 1973, one government official estimated that there are about 240 billion barrels of oil off the eastern continental shelf of the United States. That's a lot of oil. If recoverable, it would amount to an approximate forty-year supply at our present rate of consumption. Later, in conversations with oil company executives, we obtained estimates ranging from five to forty billion barrels. The truth of the matter is that nobody knows. Most of the estimates are based on analogy with other oil-producing sites sharing certain characteristics in common with the outer continental shelf. At worst, these estimates represent wild speculative guesses; at their best, they also represent wild speculative guesses.

But why do these guesses differ by so much? If you were to accept the government official's estimate, you would have to agree with Ralph Nader (he has a statistic for *everything*) that we are drowning in oil. If you accept the least optimistic oil company estimates, you must wonder if it is worth the trouble and the financial risk to go after the gooey stuff. When you look a little deeper, you realize that there are two basic reasons for wide disparities in the estimate of "oil out there." One involves definition and the second is motivation.

What do we mean by oil off the outer continental shelf? One definition is "all the oil that is presumed to be out there somewhere." This is a very convenient definition for the government official to use because he is hoping to sell those leases to the oil companies (motivation). A figure of over 200 billion barrels sounds like something worthwhile going after. But the oil executive rules out this definition. He's going to have to pay the leases. It will break his consumer-oriented heart if he must pass on the increased cost to the person to whom he trusts his car. He says, "There's no way we will ever find all of that oil under the ocean even if it's there. Some will be too deep; some will be in small deposits that we'll never locate; some will not be economically recoverable even if we find it. Why, even the best land-based wells rarely yield more than thirty percent of their contents." Quite naturally, the oil company exec-

table 3.2 How a government official, oil industry representatives, and an environmentalist can arrive at widely discrepant disembodied estimates of oil off the eastern continental shelf. Different definitions involve: Total oil, oil that can be located, oil that can be recovered at reasonable (by who's reason?) economic cost.

Estimator	Estimates
Government Oil Official	240 billion barrels of oil off the Atlantic continental shelf. Enough to last our nation 40 years.
Optimistic Oil Official	We'll find half of it (120 billion barrels) and recover one-third. That gives us 36 billion barrels, or enough to last our nation 6 years.
Pessimistic Oil Official	We'll find one-fourth of it (60 billion barrels) and recover 25% of what we discover. That gives us 15 billion barrels at the upper limit. But we think there are only 118.5 billion barrels to begin with. That means we can expect to recover only 7.40625 billion barrels. That may last us 1.1901 years or 14.28 months.
Environmentalist	They'll find one-tenth of it, at great cost to the environment. Of the 24 billion found, they'll recover 20%. That's less than 5 billion barrels of recoverable oil. Not even one year's supply.

utive plays the part of the reluctant bridegroom. He wants to get the best possible deal on oil leases. But even he must be careful to play the numbers game with consummate care. If he sets too low an estimate, the environmentalists will clobber him in the courts. They will show that the desecration of the seascape is too high a price to pay for so little oil. Table 3.2 shows how a whole lot of oil can be either a whole lot of oil or a mere dribble, depending on which definitions of "oil out there" you are willing to take.

No wonder the American public, the Congress of the United States, and the person who pumps gas at the corner station are confused by the energy crisis, the population explosion, the crime rate, additives in food, the environmental crisis, resource shortages, and the thousand and one (notice my precision—it feels so good) issues that vie for our attention these days.

Now, having shown that the disembodied statistic can apply to such unimportant things as energy and the environment, let us return to vital issues like clean and pure soap. What exactly have we learned when we are told that something is 99.44 percent pure?

First, I guess, we had better find out what pure means. In the dictionary I consulted, I found over twenty different definitions including: a mixture —as of dogs, or pigeons' dung in water, for bating kips and skins after liming. I haven't the foggiest notion of what a kip and liming are and I am too lazy to look them up. And bating sounds obscene. However, I doubt that pure was being used in this sense. But let us stipulate that we have arrived at a satisfactory definition of pure. What does 99.44 percent pure tell us? How does the purity of this soap compare with that of competing brands? For all we know, 99.44 percent may be a rather low level of purity. Maybe a Brillo pad is 99.86 percent pure. Should I then wash my face with Brillo? And while we are on it, what does purity have to do with the cleansing powers of soap? Maybe a soap that is 52.53 percent pure does a better job of cleaning. I don't know and certainly the ad doesn't help me decide.

Nor was I told much when Volvo informed me that 90 percent of their cars sold in America over the past ten years are still on the road. (However, I suspect that very few Volvos were sold here ten years ago. My guess is that most U.S. Volvos are of rather recent vintage. It would be a supreme disgrace if nearly 100 percent of these are not still in service.) Nor when I see that graph showing that Bufferin (or is it Anacin or Excedrin? I get them all mixed up), with a higher level of pain reliever, gets into the bloodstream faster and maintains a higher level of pain reliever longer than competing brands. Why not just take more of the chainstore brand of aspirin and enjoy the same high level of relief at a lower cost?

Do you want to bring fun back into TV viewing? I recommend you forget about the programmed stuff and concentrate your attention on the commercials. But don't be offended when you suddenly realize that someone out there thinks you have the intelligence of a three-year-old chimp.

Chapter Four: Bias, Bias, Everywhere

A great deal of statistical information is gathered in a special setting referred to as an experiment. In an experiment, there are a minimum of two groups of subjects. These groups are treated alike in all respects except for the administration of the experimental treatment. Various data are collected and then analyzed in terms of various descriptive statistical techniques. Then inferential statistics moves into the spotlight and tells us whether the observed differences in our sample

statistics are of such a magnitude that we can rule out "chance variation" as the cause of the differences. In short, inferential statistics permits us to draw conclusions concerning the *effects* of experimental variables and to generalize these conclusions to broad populations.

A key element in any research is the strategy used to rule out the possibility that unsuspected variables will "contaminate" our observations. For example, some years ago it was reported that plants that were "prayed over" fared far better than control plants that received only the usual TLC. Now I have nothing against prayer and would welcome proof that prayer assists plant propagation. (Personally, I sing to my plants. I'm a great believer in shock therapy.) I cannot ignore the fact that prayers frequently issue from the mouth. And breath is high in carbon dioxide, moisture, and warmth—all factors that favor the growth of plants.

In this chapter we look at some of the psychological factors that might inadvertently breathe moist air into our research program.

Liars, All of Us

If the truth be known, it is probable that each and every one of us, saints and sinners alike, has told thousands of lies in a lifetime. (How's that for a disembodied something or other?) And I am not talking about the purposeful effort to deceive. Research in the field of psychology has made it abundantly clear that our motivations and our expectations profoundly influence the way in which we look at the world: what we pay attention to, how we perceive the things we see and the events we attend, and how we remember and report what we have previously seen. Let me illustrate this by an example from my own experience.

During World War II, I was assigned to the Third Army (George C. Scott's Army, you may recall) as a medical aide man. Over the course of a number of different campaigns in Europe, I was placed on temporary duty to many different organizations, some of them including front-line outfits where my job was to administer emergency treatment to GIs wounded on the battlefields. During those stints, I was referred to as a medic.

After the great thrust across France, the U.S. forces became stalled along the entire Siegfried line. If you don't know about the Siegfried line, let me tell you something about it. It was an enormous system of bunkers constructed out of steel-reinforced concrete as much as ten feet thick. Efforts to penetrate this barrier were constantly frustrated. Ninety-mm shells from U.S. tanks just bounced off the walls and accomplished little more than giving the bunkers a pock-marked appearance. About the only system that could penetrate the mausoleums was dangerous, laborious, and time-consuming. A flame thrower was directed at each bunker to force closure of the observation window, thereby permitting our movements to go undetected. As soon as the bunker lost its eyes, an explosives expert would rush up to it and attach a charge of explosive plastic to the bunker. With what seemed like millions of these pill boxes in the way, it looked as if it would take months or even years to clear out a sufficient number to permit penetration into Germany.

There had been rumors all along the front that the Air Corps would come to our rescue by sending over squadrons of flying fortresses to drop blockbusters on these bunkers. One morning as I looked out of my foxhole, I saw a tremendous number of B-17s or flying forts moving over us toward the German line. Giving credibility to the rumors, they began to drop house-size bombs on these bunkers. The Germans, of course, replied. They sent up their usual ack-ack and tried to discourage the B-17s by a concentrated firing of a varied assortment of anti-aircraft weapons. Suddenly I saw one of these B-17s struck in the underbelly with what was obviously a mortal blow. It began to move in that sort of sliding, sideward motion characteristic of a plane that has been hit and is going eventually into a terminal nosedive. Without thinking of the obvious dangers, I jumped out of my foxhole and began to shout "Bail out, bail out," knowing full well that the B-17s included, I believe, a crew of ten. Don't ask me why I did it. It was a silly thing to do. Not only did it expose me to possible enemy fire but, even if by some miracle the occupants of that B-17 could hear my shouted exhortations, it is altogether probable that the same idea had occurred to them. In the adjoining foxholes other American troops looked up and saw the same floundering B-17 and joined me in the chorus of "Bail out, bail out." Before long the entire front in my vicinity, perhaps involving as many as 50 to 100 battle-weary foot soldiers, were out of their foxholes and screaming, "Bail out, bail out." The amazing thing is that during this time we received absolutely no fire from the German side. Perhaps it was because they were out of their foxholes and screaming, "Don't bail out, don't bail out." I really don't know. The point of the matter is that we all engaged in this very silly behavior of shouting to individuals who couldn't possibly hear us.

The plane continued to slide into its appointment with oblivion. My vision of the "fort" was so clear that I could even see the door from which the crew would bail out. But bail out they did not. There was absolutely no sign of any motion whatsoever on the exterior portals of the plane.

I found myself overcome with a feeling of crushing irrelevancy at the realization that I could do nothing to alter the inevitable. All I could do was scream louder and louder as the B-17 plummeted closer and closer to the earth. Then, suddenly we perceived a vagrant scrap of tinfoil floating over our heads. In retrospect, it is hard to believe that a piece of tinfoil was what we had perceived as a B-17 disabled by enemy fire. The truth of the matter is that my own fears had greatly influenced the nature of my perception of the realities

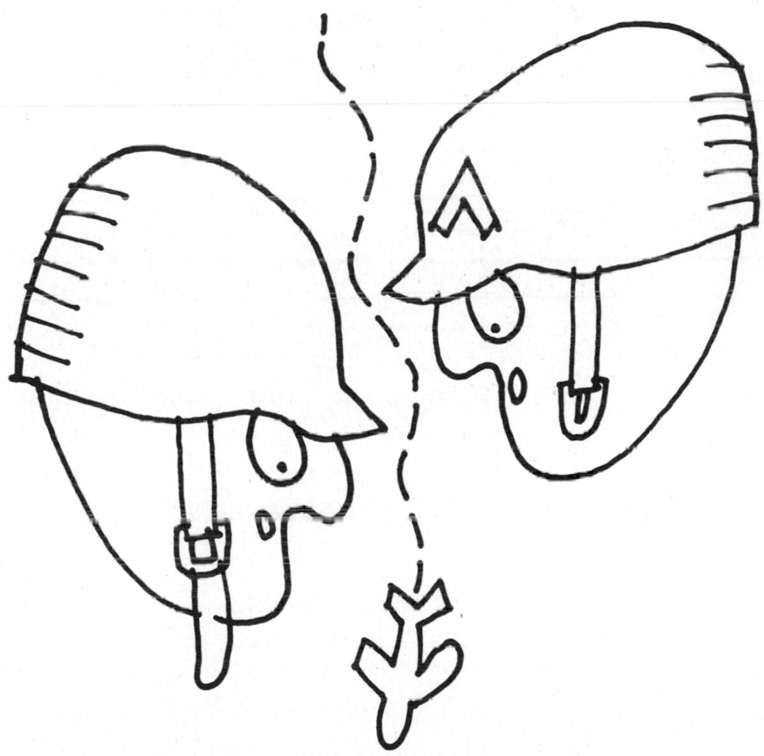

of the situation. The tinfoil *could* have been a plane in distress. My own motivations transformed what could have been into something that was. But my fears were shared by the dozens of other front-line troops. That's why so many of them looked up, saw a silvery sliver sliding across the vastness of space, an ambiguous stimulus transformed by their fears into a B-17 in distress.

Now you may wonder how that piece of tinfoil got there. In World War II, the radar was not nearly as sophisticated as it is now and one of the means of jamming enemy radar was to release huge quantities of tinfoil when you got near the areas where radar might be sending out its electronic probes. And obviously it was our own B-17s that dropped this tinfoil in order to jam enemy radar. But the point is that, in this instance, I was telling a lie—an inadvertent lie, to be sure, but still not the truth. In effect, I told all of those other fellows on the front line, "Look, there is a plane; one of our planes in great trouble; it is falling" (shades of Chicken Little). This lie was believed by a whole bunch of people, a lie not told with malice aforethought or desire to deceive. It was simply a case in which I had failed to perceive reality as it was because my motivational system had overwhelmed my perception and robbed it of objectivity.

Now let's take a look at another story, this one from contemporary psychological research. A psychologist by the name of Martin Orne has done a great deal of research on a phenomenon he has called "the demand characteristics of the experimental situation." In summary, Orne's research has led him to a rather startling observation: In studies using human subjects, there is a strong possibility that the outcome of the experiment may be influenced or even determined by the ways in which the subjects view the purpose of the study. In other words, the results of experimental studies may reflect the subject's perceptions of the study rather than the effects of the experimental treatments or manipulations. Let's elaborate on the types of observation that led Orne to this troublesome conclusion.

Orne's reasoning started out with the view that in our society an experiment has a very special meaning. Most people agree that experiments are of importance and lead to results that are ultimately for the betterment of our species. As long as experiments deal with inanimate objects in nature (such as in physics and most branches of chemistry), there are no special complications. However, when the objects of the scientist's investigations are human, a whole host of interesting and provocative problems arise. The human subjects involved in a study assume that they are participating in something of importance. The experimenter obviously feels that he or she is also doing something of importance; otherwise he or she would not be conducting the research. The result is a belief, bordering sometimes on reverence, shared by both experimenter and by subject to the effect that their joint efforts will lead to significant findings. In effect, Orne argues that an experiment using human subjects involves an implicit contract between experimental subject and experimenter. Each one is expected to perform a certain role. By tacit agreement, the experimenter is acknowledged as a person of authority and it is felt that what he or she demands of the subject should be carried out by the subject obediently and frequently without question. On the other hand, the subject feels that he or she must cooperate in all ways possible so as to make a complete success of this joint venture.

To illustrate the extent to which people in our culture regard an experiment as something very special, Orne conducted an informal study with a group of his own acquaintances. He approached them and asked if they *would do him a favor*. When they agreed, he asked them to do a set of five pushups. These acquaintances, as you can imagine, responded with a certain degree of incredulity. They demanded an explanation. Orne repeated this observation with a similar group of individuals at another time. However, the phrasing of the request was just slightly varied. This time Orne asked if they would mind *participating in an experiment* of brief duration. Upon their agreement to do so, he asked them to perform five pushups just as he had asked the previous group to do. However, the response

I'd feel silly if this weren't an experiment.

was different this time. Instead of asking "Why?", they asked "Where?" It is clear that most subjects want to participate in experimental investigations. They want to perform their roles as subjects well so that scientific knowledge can be advanced. This desire is frequently attested to by a question commonly raised by the subject on completion of a study: "Did I mess up the experiment?" This is their way of expressing fear that they might not have performed their role as subjects in an appropriate manner.

Orne believes, and there is now a body of experimental evidence to support this view, that there is a constant probing by the subject in an effort to find out what sort of behavior is expected by the experimenter. As you may or may not know, a great deal of psychological research involves either deceiving the subject as to the purpose of the study or merely leaving him or her uninformed until the study is completed. In an ambiguous situation of this sort, subjects put their own interpretation on the purpose, the goals, and the expectations of the experimenter.

Subjects attempt to perform in ways that are consonant with the interpretations they have made. If they feel the experimenter is looking for data of a certain sort, they will modify their own behavior so as to provide the type of information they believe the experimenter desires. Indeed, for their part, scientists are not the cold, aloof, objective, dispassionate, unfeeling, bloodless creatures that are often depicted in fiction. They are made of blood, muscle, gut, sinew, and bone, just like the rest of humanity. They also want desperately and feverishly to be right. They want to have their hypotheses confirmed just as much as an automobile salesman wants to close sales, just as much as in insurance salesman wants to have you increase your coverage to

$200,000, just as the impoverished oil company executive wants to see his company finish the year with a better than 10-percent profit on capital investment. In short, experimenters are not bloodless organisms. So there is a real possibility that they will inadvertently communicate to the subjects—through body language, or a smile, or a nod, or the tone of voice—precisely the type of behavior in which they want the subjects to engage. So the experimental setting, usually conceived of as a sterile and formal transaction between experimenter and subject, becomes, under Orne's interpretation, a seething cauldron of unspoken and unconscious contracts.

Let us take a moment to review what we have been sharing. In an experiment, the subject feels that he or she is participating in something of importance; the experimenter also feels precisely the same way. Both of them are playing roles. The experimenter has the role of the investigator, the God-surrogate who is trying to arrive at truth. The subject plays a role in which he or she is willing to be subjugated to the purposes of the experimenter. The goal of both actors is to make a contribution to our knowledge of the human condition. Both the experimenter and the subject want to succeed in what they are doing. The subject is usually uninformed about the purpose of the experiment and is constantly looking for hints from the experimenter as to what is going on. The experimenter may inadvertently communicate to the subject what he or she wants the subject to do. Under these circumstances, the results of experimentation can lead to an accumulated body of scientific fiction. Orne suggests that the results of many experimental studies, rather than constituting an objective examination of some experimental hypothesis, may actually reflect the success in the subject's understanding of what is expected of him or her, that is, the demand characteristics of the experiment. To the extent that Orne is correct, a whole body of experimental research in the fields of psychology, sociology, and economics must be regarded with an air of healthy skepticism.

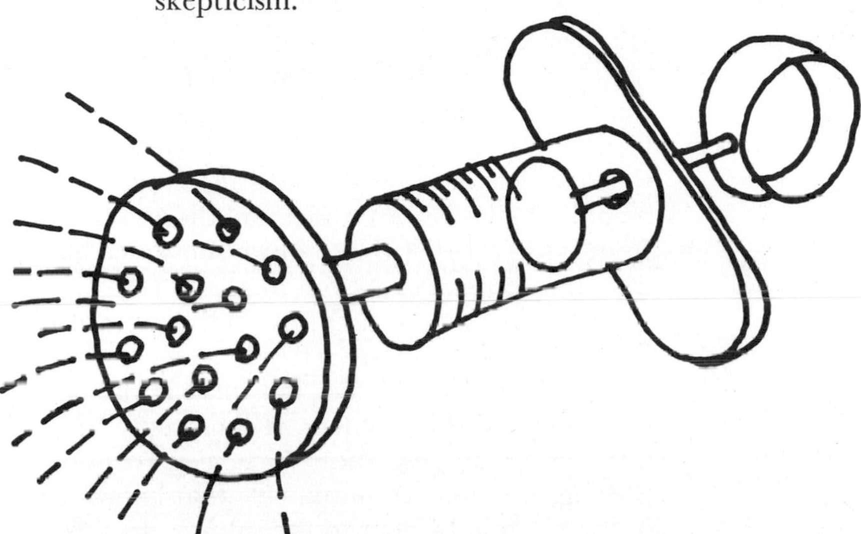

In recent years, a number of efforts have been made to overcome this problem. For example there is an increasing use of what is called the "double-blind" study, in which neither the subject nor the experimenter knows what treatment is being administered. Let me just briefly explain this type of experimental design by reviewing the original, large-scale Salk vaccine study. Many of you may recall this study even though you must think back a couple of decades in time. It had one most unusual characteristic: The subjects were never told whether they were getting the vaccine or the placebo (a substance that is identical in appearance with the vaccine but has no known effect on the body). But the Salk study went beyond this. Not only were the subjects uninformed as to whether they were getting the vaccine, but the individuals administering the vaccine were equally uninformed. Indeed, not even the doctors who ultimately had to make the diagnosis of suspected cases of polio were told which subjects received the vaccine and which the placebo.

Let's see how this double-blind analysis works. It is well known that many ill people will get better so long as they are convinced that the treatment is going to cure them: Sugar-water can be as effective as the latest miracle drug if they think the sugar-water is that drug. Improvements in behavior or physical condition resulting from faith or belief that a given substance has curative powers is referred to as the *placebo effect*. By keeping the subjects in the blind as to identity of the substance they have received, the placebo effect is defeated. But there is also a second side to the problem. If the people administering the drug know what they are administering, they may communicate that knowledge in some subtle fashion to the subject. In this event the subject would know whether he or she has been administered either the placebo or the drug. The final step in the Salk study was to keep the family doctor equally in the dark. Why? The doctor had to make the diagnosis. It was felt that if the family doctors had knowledge of the treatment received by their patients, their diagnostic judgment might be subtly influenced. How could this be? The fact is that polio was often confused with the common cold. Many individuals would contract what would appear to be a summer cold; they would have no symptoms other than a sore throat and a runny nose and they would recover in a week if treated and in seven days if not. For all intents and purposes, polio episodes that produced nothing more than cold symptoms were no better or worse than a cold. The truth of the matter is that very few people who contracted polio displayed the crippling effects that we see in the very few unfortunates who were paralyzed by the polio virus.

Now a doctor who knew that his or her patient had received the vaccine might very well have diagnosed an apparent summer cold as a summer cold. On the other hand, if he knew that the patient received a placebo, he might have diagnosed that summer cold as polio. This is not suggesting dishonesty on the part of the doctor

---

box 4.1 *Patent medicines and placebo effects*

Have you ever had the good fortune to find an old snake oil bottle complete with original label? I am fortunate to have in my bottle collection several outstanding examples. My favorite is "The Wonderful Japanese Oil." The label on the bottle reads:

> *For Rheumatism, Neuralgia, Spinal Complaint, Headache, Toothache, Cramps, Sprains, Bruises, Contusions, Chapped Hands, Burns, Chilblains, Fever Sores, Sore Throat, Swelling of the Tonsils, Kidney Complaint, etc., to be used externally on the parts affected, once or twice per day. For severe cases of Rheumatism, take ten (10) drops inwardly, night and morning. For Summer Complaint, Cholera Morbus, Cramps in the Stomach, etc., for persons under six years of age five (5) drops; over six years eight (8) to ten (10) drops, inwardly, in a little sweetened water. This Oil cures Rattle Snake and other Snake Bites.*

I'll guarantee that any medicine that can cure so many things has a powerful placebo effect going for it. Or it contains so much alcohol that it stones people out of their pain.

---

but is only giving recognition to the pervasive influence of the faith of most doctors in medical science. If the doctor believed that the polio vaccine would work, he or she would be strongly influenced to interpret the results in the direction that would confirm that belief.

Now let's apply the observation that we have made in this chapter to an area that is one of my favorite whipping boys: media advertising. How many times have you seen an advertisement on television in which an individual, filmed by hidden camera, is given a choice between two products? Invariably the discriminating housewife or the harassed businessman selects the product that is being pushed by the commercial. What is left unsaid is that only about one-half of the people filmed ever appear on the commercial. You see, the other half made the unforgivable error of selecting brand X. No advertiser in his right mind would permit the unfavorable commercial to be shown. The result is a commanding appearance of unanimous endorsement of the sponsor's product. Note that the deception is one of omission rather than of commission.

The whole sordid affair is in many ways reminiscent of a con game that was prevalent many centuries ago in Europe. A group of enterprising individuals claimed that they could control the sex of the child of pregnant women and they were willing to put their money where their mouth was. If a woman wanted to have a boy, for example, she was given an ironclad guarantee that her child would be a boy. Of course, to receive this guarantee she had to pay a sum of money. For example, if she paid 50 livres, she would be assured of the delivery of a bouncing baby boy. But what if by some rare quirk of fate a girl was born? "Lady, that can never happen," they assured the expectant mother. "But to prove our integrity and absolute faith in our proven method, we will return your money and give you an extra 35 livres if we're wrong. So I ask you, madame, what can you lose? If our method works, as we know it will, you have your boy. If God, in His mysterious ways should decide otherwise, you gain 35 livres. You have nothing to lose but everything to gain."

Now assuming that there were 100 takers, the income was 100 × 50 livres, or a gross income of 5000 livres. But, on the average, about fifty of the births would have been an embarrassment (not much, I assure you) to the entrepreneurs. That meant 85 × 50 livres would be refunded. That comes to 4250 livres. Net profit? 750 livres. Certainly beats working for a living.

The same applies to TV advertising. The latest gimmick is to have claims certified by an independent research organization. The scenario goes like this. A given manufacturer has placed its advertising account with a particular agency. The agency would like to have objective and independent proof that the product is superior to a competitor's ware. Consequently, it hires an independent research firm to collect and analyze data and report on its findings. Everything is strictly aboveboard. There is no attempt on the part of the advertising agency to influence the findings of the research organization. Imagine, for example, that the study is sponsored by Gasguz Automobiles. It is hoped that your agency can demonstrate that many people find that Gasguz provides a smoother, better ride, with handling characteristics superior to those of the more expensive competitive lemon. But if you want to be sure that you conform to the truth in advertising requirements of the FCC, you had better make certain that this comparison will not expose you to litigation for fraudulent advertising practices. In other words, you cannot go to an independent research firm and say, "Now here are Gasguz automobiles and here's brand Y. We want you to prove that Gasguz automobiles are better than brand Y." No, certainly not. But—and this is a big but—Gasguz does pay the bills. It also provides a nice stipend to be paid to each subject who volunteers to participate in the study. Often the subject is given a choice of contributing to his or her favorite charity or accepting the money for personal use. (Some years ago I conducted several consumer panels involving Lincoln Continental and Cadillac owners. My major observation was that Cadillac owners accepted the cash, while the owners of Lincoln Continentals contributed the money to charity.)

Now let me assure you of something. No concerted effort is made to conceal the identity of the manufacturer who is footing the bill. The researcher surely knows and it doesn't take the subject long to find out. For, as we know, each subject is constantly sending out feelers in an effort to ascertain what the experimenter is hoping to find. And this probing need not be conscious and calculated. It need not be a situation in which the subject thinks, "I am purposely attempting to provide erroneous information so as to satisfy the experimenter." No, it's a much more subtle type of transaction that takes place, and it could be completely at the unconscious level. The individual is playing a role, that is, to provide desired or necessary information concerning the relative features of these two automobiles. As we previously concluded, the researcher knows who is paying his or her fee, and quite naturally would like to find that most people prefer the handling characteristics of Gasguz automobiles to the more expensive competitor. The subject is probing to find out what the researcher wants. So the circumstances are ideal for the demand characteristics to assert themselves in the transactions between researcher and subject. To the extent that the subject is successful in inferring the motives of the experimenter and to the extent that the experimenter is successful in communicating what he or she wants, even if completely inadvertently, the results will favor Gasguz over brand X, Y, or Z or any other brand. I could easily imagine circumstances in which the Tin Lizzy could be found by "an objective and independent" appraisal to be clearly superior to Rolls Royce.

Let us hope that the independent research firm never discovers the double-blind technique. Otherwise, we may be forced by the weight of evidence to swallow that junk they feed us during the commercial break.

NATURAL
NATURELLE
NEW!
FREE SPOON INSIDE
SEND FOR ANDREW WYETH MAKE-UP KIT

Chapter Five: Gaffing with the Graphic

If you want to land someone hook, line, and sinker, gaff him with a gorgeous graphic. For my part, I'll take it over the centerfold of *Playboy* any day of the week (I also lie a lot). There is something almost hypnotic about a well-prepared graphic. It commands your attention and often benumbs your critical faculties. Particularly seductive are the animated ones shown on television with the lines that really move. They blow the mind.

Now don't get me wrong. Graphics are the lifeblood of the descriptive aspect of statistics. Just as a picture is worth a thousand words, so is a well-constructed graphic worth a thousand data points. There is a veritable wealth of information that can be conveyed with a few lines, a bar or two, or a circle divided up like a pie (cut the pie in half if you want to, but give me the bigger half).

Unfortunately, the very virtues of graphics are their Achilles heel. They can be attractive, particularly when executed by an artist with a flare for the unusual. However, there is always someone poised on the sidelines waiting to peddle the pulchritude in the bawdy houses of commerce. Like the skilled stage magician, the artist can direct your attention to one part of the graph in order to distract you from another part where monkey business is going on. Perhaps the best example of graphic legerdemain is in the type of deception I call the "validating pseudograph."

Validating Pseudograph

You've seen this one on TV hundreds of times, and so have I. A person is providing expert testimony on the efficiency of some product that he or she is hawking on TV. In a graph shown to the side, the lines move in such a way that they appear to prove the validity of the testimony. Great care is taken never to label the axes. For example, Fig. 5.1 discusses a hypothetical engine additive that promises to add new life to the geriatric set of gas guzzlers.

fig. 5.1

Actually, this type of deception doesn't bother me too much. It is so obviously dishonest that I can sit back and enjoy its ineptitude, in the same way that I occasionally enjoy a grade-B movie or the usual TV fare.

But "rubber band boundaries" are something else again.

The Rubber Band Boundary

What makes the "rubber band boundary" chart particularly insidious is that it is generally used to present important data and commonly appears in serious publications (I have seen them in the most highly respected newspapers as well as in weekly news magazines). The rubber band boundary chart is really the progeny of graphic anarchy. By this I mean that there are no universally agreed-upon methods of representing the relative lengths of the vertical and horizontal axes. Therefore, these axes are like rubber bands, ready to expand or contract on demand of the user.

Let us say that the user is a sales manager of a department in a large discount store. He has kept a careful record of the sales figures of all the personnel working under him and he wants to use these figures to put a burr under the saddle of salespeople who, he feels, are not giving their all to the dear old shop. What does he do? He draws bar graphs representing the weekly dollar volume for each salesperson but he stretches the vertical axis and contracts the horizontal. He states, "Ms. Dee, you are obviously not holding up your end of the work load. The rest of the department is carrying you. If you don't believe me, look at the sales chart." (See Fig. 5.2a).

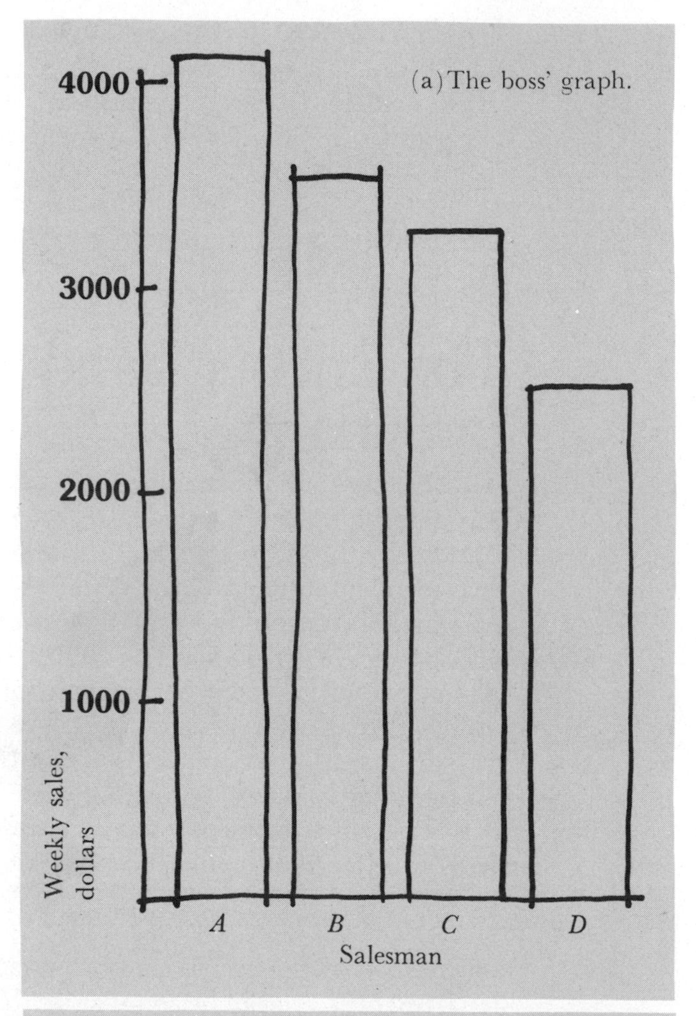

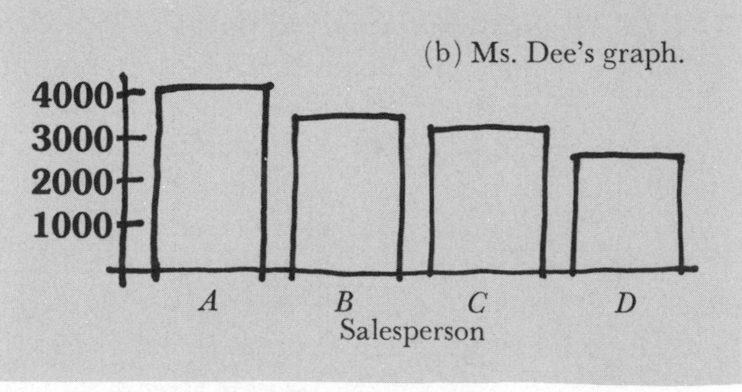

fig. 5.2

The rubber band boundary chart. Markedly different impressions can be given by alternately stretching and contracting the axes.

But Ms. Dee has not been caught off guard. Without her manager knowing it, she has been taking prolonged coffee breaks each afternoon while preparing the hors d'oeuvre shown in Fig. 5.2(b). She replies, "It's strange to hear you say that. As you can see, I'm doing about as well as anybody. A few sales, one way or the other, and I would be the top in the department."

It is because of the elastic axes that I have long advocated the three-quarter high rule as a means of ending graphic anarchy. This rule is expressed as follows: "The vertical axis should be laid out so that the height of the maximum point is approximately equal to three-quarters of the length of the horizontal axis." * The rule has the virtue of removing personal preferences and biases from the graphic decision-making process. Figure 5.3 shows the sales data of the previous example presented in accordance with the three-quarter-high rule.

Weekly sales figures shown in accordance with the three-quarter-high rule.

fig. 5.3

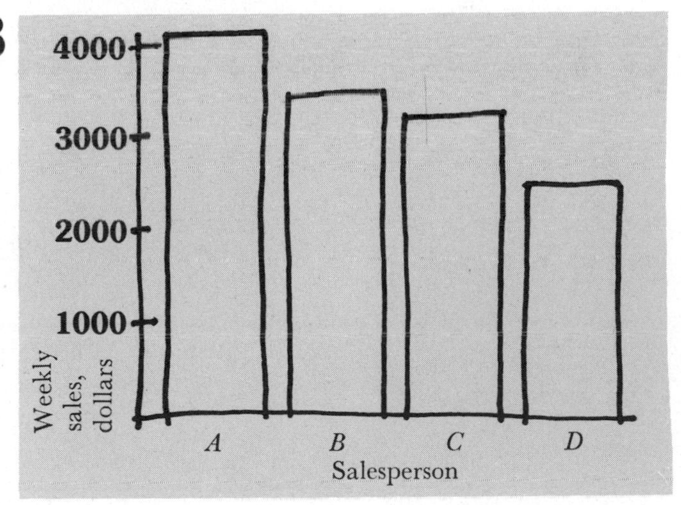

* Runyon, R. P., and A. Haber (1976). *Fundamentals of Behavioral Statistics,* 3rd ed. Reading, Mass.: Addison-Wesley.

But, if you think the rubber band boundary chart is bad, wait until you see the "oh boy! chart."

Oh Boy! Chart

All you need to qualify for knighthood in the "Order of the Oh Boy! Chart" is to forget that most things have a zero point. If you want to exaggerate the rise or fall over time of some data points, you merely begin the vertical axis with some numerical value other than zero. The value you select depends on just how unscrupulous you are. Let us say that you're a stock-market analyst who puts out a weekly newsletter for which your subscribers pay a pretty price. During the week of 24 November 1975 you warned, "Look for a greater-than-average drop in market prices because of instability over the New York bankruptcy situation, continued indications that the Arab nations will be looking for higher prices

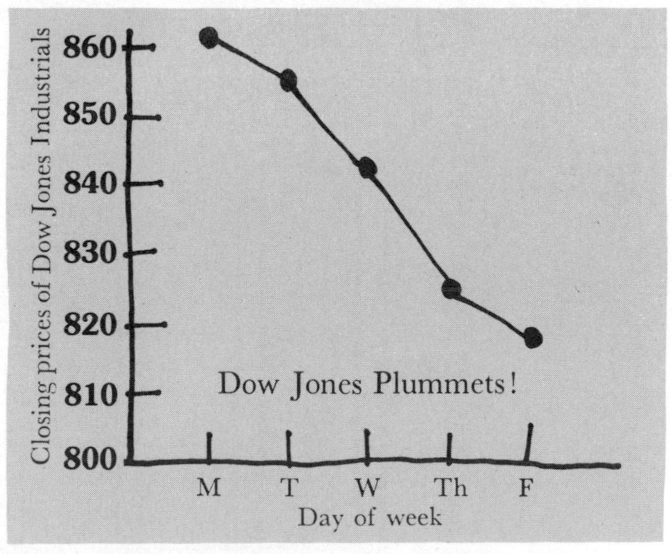

fig. 5.4

Oh Boy! chart. The falling trend is exaggerated by failing to show the total range of possible closing prices starting at zero. It is as if one held a magnifying glass to only one part of the total graph. (Based on the Dow Jones closing averages in the New York Stock Exchange during the week of 1 December 1975.)

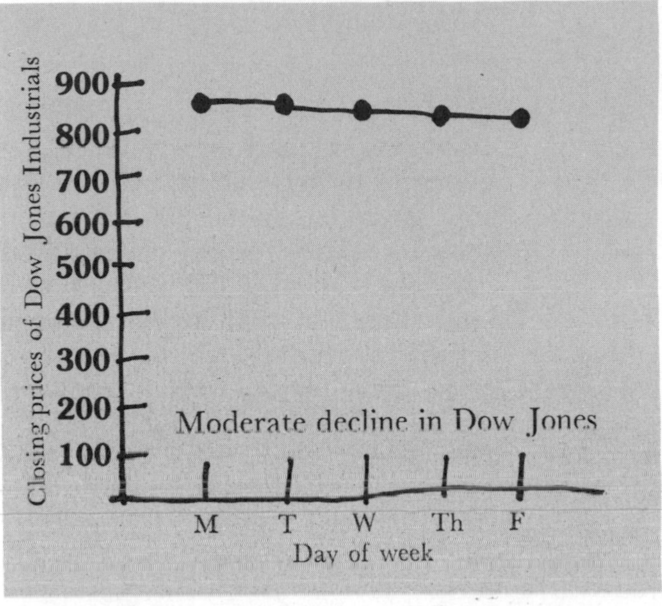

fig. 5.5

Graph of closing prices of Dow Jones Industrials during the week of 1 December 1975. Note that the full range of possible declines is shown.

in oil, the civil strife in Lebanon, and economic indicators that show we are not emerging from the recession as fast as we would like." You then show the chart of Fig. 5.4 as proof positive of your prophetic abilities.

Your competitor also puts out a newsletter. In it, the following prediction was made: "Look for relatively stable or moderately falling market prices because of mixed indicators; on the one hand, there is some hope that the federal government will intervene in the situation in New York City, and continued indications that the Arab nations will be taking a less extreme view concerning oil pricing policies. On the other hand the civil strife continues in Lebanon, and various economic indicators show we are not emerging from the recession as fast as we would like." As proof of this astute economic forecasting, your competitor produces the chart of Fig. 5.5.

What's that? You are offended by the second chart? Why?

"Well," you say, "by showing the whole smear of possible values from zero to 870, you are obscuring a drop that actually took place. Using your method, even a drop of 100 points would show up as only a minor perturbation. Believe me, baby, a drop of 100 points in a week would send a bunch of small investors scurrying pell-mell for the gas pipe."

You've got a point there, I must admit. But if you're going to direct attention to only a limited part of the total range, the least you can do is show a distinct separation at the bottom of the vertical axis so people will know what you have done. And don't label the chart "Dow Jones Plummets." Credit your readers with sufficient intelligence to judge for themselves. (See Fig. 5.6.)

Pop-Eyed Pictographs

Some people find bar graphs and line graphs exceedingly static, unimaginative, and boring. Instead of bars, they substitute pictures of the things represented in the graphs. Such a figure is referred to as a pictograph or a pictogram. The Bureau of the Census is fond of using human figures to represent population statistics (see Fig. 5.7). (Parenthetically, they might be accused of some insensitivity since the last figure to the right is almost always seriously maimed, lacking, as he or she does, various parts of the body.)

Pictographs are perfectly legitimate ways of displaying statistical facts. They can be both interesting and, if done correctly, informative. But there's always someone trying to improve on Mother Nature. So instead of using hundreds of tiny little human figures, our budding genius

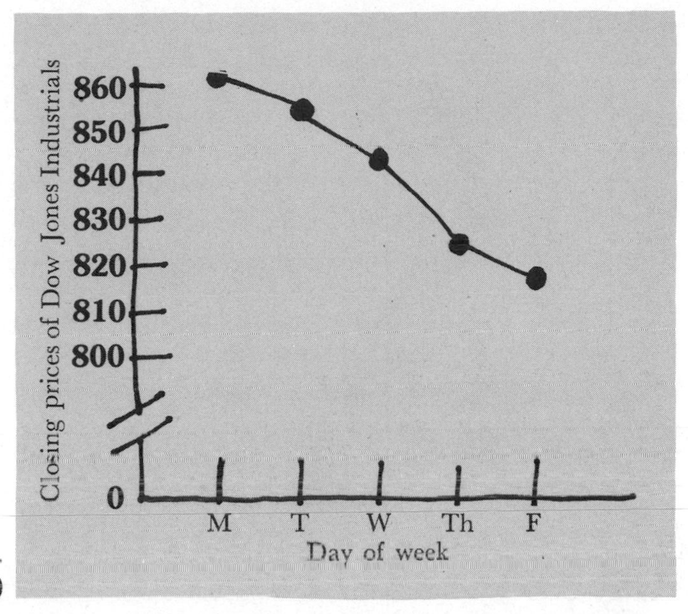

fig. 5.6

If you must use an Oh Boy! chart, at least let your readers know that you have taken some liberties with the vertical axis.

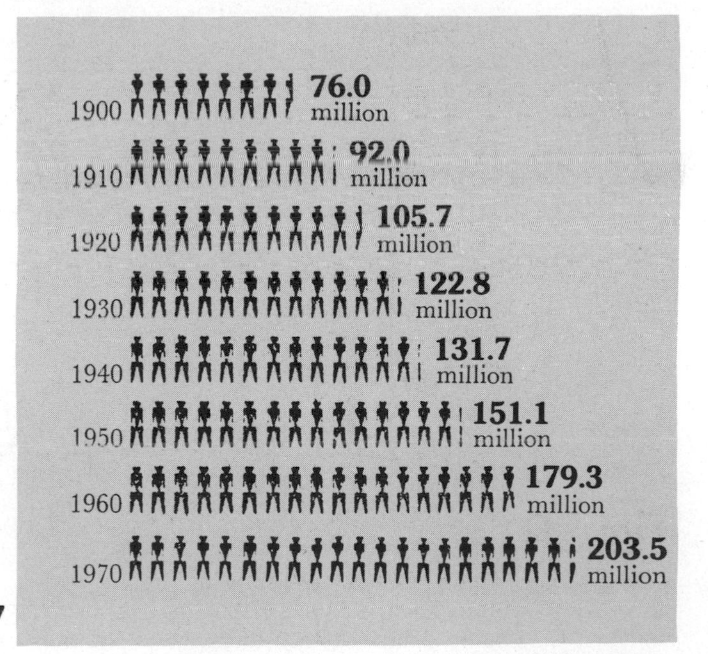

fig. 5.7

Population of the United States from 1900 to 1970. Each figure represents 10,000,000 people.

fig. 5.8

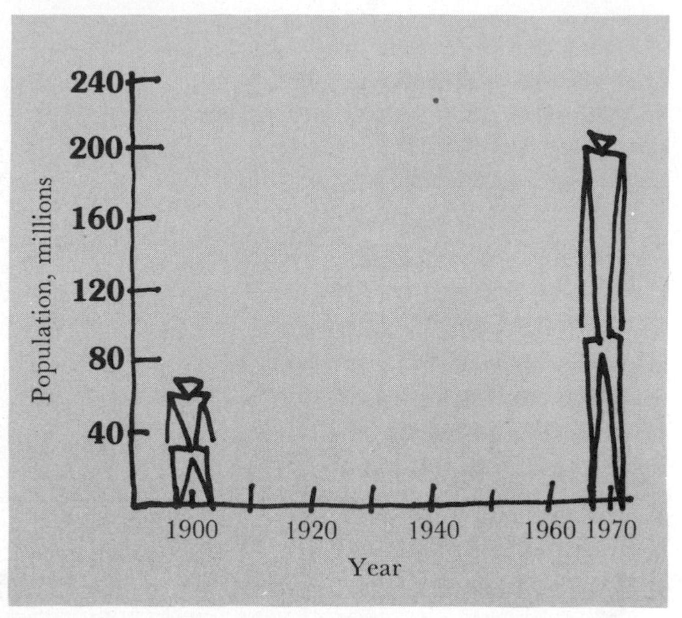

Pictograph showing the population of the United States from 1900 to 1970.

fig. 5.9

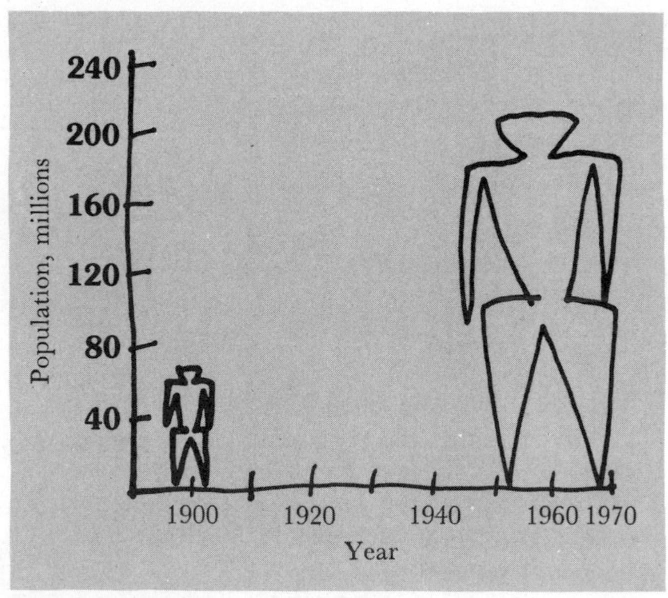

Pop-eyed pictograph showing the population of the United States from 1900 to 1970. Although the *height* of the figure is meant to represent the population, most individuals respond to the *total volume* of the figure. Thus, the difference in growth is grossly exaggerated.

reasons, "I'll draw just one figure but I'll make it proportional to the total. Serendipitously, I'll also avoid the dissected human figure." The product of his cerebrations is displayed in Fig. 5.8.

Our friend ponders over his artistic production. "Something seems wrong," he observes. "Why, of course. The human figures are distorted all out of their usual proportions. That poor creature on the right must have spent his last three weeks on the rack. I must restore the proper proportions to the human anatomy." He draws Fig. 5.9 and is very happy this time with his artistic endeavors. "Not only is it more pleasing to the eye," he observes, "but it somehow makes the growth in population seem even more substantial." Well, of course it does. He has not only improved the proportion of the human figure, but he has vastly increased the total volume.

Granted that the *height* of the figure is supposed to convey the information about the population, it is difficult to ignore the overall "King Kong" proportions of the figure to the right. Such figures defeat the basic purposes of honest graphic representation—to convey information interestingly, rapidly, and accurately.

The Double-Whammy Graph

The following type of graph is sheer delight for the perennial prophets of doom. Take a good look at Fig. 5.10. It tells a two-fold tale of grief and despair. First, there has been a steady erosion of the purchasing power of the dollar since 1950. "A dollar today just ain't worth a dollar any more," laments the modern-day Jeremiahs (incidentally, I wonder when a dollar was really worth a dollar). But if the decline of the dollar

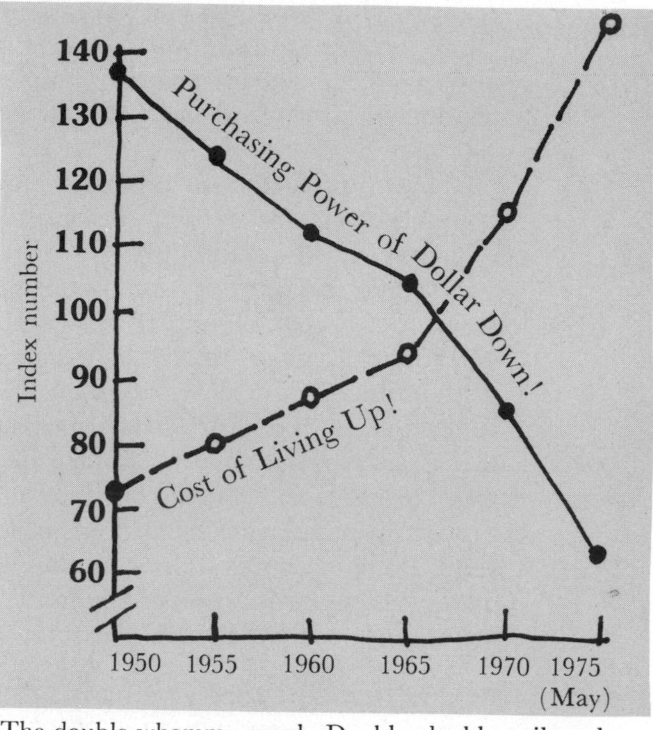

fig. 5.10

The double whammy graph. Double, double, toil, and trouble. As if inflation weren't bad enough, a second factor is added that tells the same story. The result is the appearance of two bundles of bad news. (*Source*: Data obtained from the American Almanac for 1976.)

isn't enough grief to heap on the hapless consumer, a rising curve is added to depict the inflationary spiral. The story is clear, albeit dismal. Two equally bad things are happening simultaneously. We are spending a lot more money to buy things because money is less valuable and, at the same time, the cost of that new pair of shoes keeps going up.

But wait a minute. There's something fishy here. Aren't these two measures—cost of living and purchasing power of the dollar—two aspects of the same thing, like two sides of the same coin? Of course they are. In fact, one measure is essentially the reciprocal of the other. If you do not believe me, try dividing the purchasing power of the dollar, for any given year, into one hundred. You should obtain roughly the cost of living index for the same year. Or divide the cost of living into one hundred, and you will find the value of the dollar for that year. Inflation is bad enough. Let's not compound our miseries by displaying the double-whammy graph.

The Cumulative Chart-Only One Way to Go

In this chapter we have looked at a lot of graphic skullduggery. We have seen axes pinched and squeezed like a baby's backside and stretched like a forty-five bra. We have seen line drawings masquerading as graphs and solid figures blown up like inflatable toys. Now we are going to put the icing on the cake by looking at a perfectly legitimate, extremely useful form of graphic representation that is frequently misinterpreted by both lay people and scientists alike. I am referring to the cumulative chart, a favorite graphic device of many behavioral scientists.

The cumulative chart is essentially a time/performance graph in which the performance measure is cumulated over time. To illustrate, let us say that Margaret is a clerical-typist who works under the direct supervision of her friend and counselor, State Senator Frumph. Although her job has not been clearly defined, she meets frequently with VIPs and attempts to win them over with charm and a vibrant personality. Such is her dedication to her job that she is even learning to type (some day she hopes to type the memoirs of her life that somebody else will write for her).

Senator Louie, who adores Margaret, has kept a cumulative record of the total number of words typed over the first weeks of her employment.

That is, at the end of each week, he calculates the number of words typed by Margaret and adds this figure to the total number of words she had typed over the preceding weeks. The following cumulative chart (Fig. 5.11) was the result of his efforts.

Now Senator Louie is a somewhat vain person —an occupational hazard. Surveying the chart, he muses, "My secretary's first typed words! Congrats, Louie. I guess this makes you legit as a politician. A beautiful curve. Sure have come a long way since Delancy Street." Then almost as an afterthought, he looks again: "Margaret's doing pretty good, too. That chart spells progress, no matter how you look at it. She's getting steadily better. No telling where she will end up."

fig. 5.11

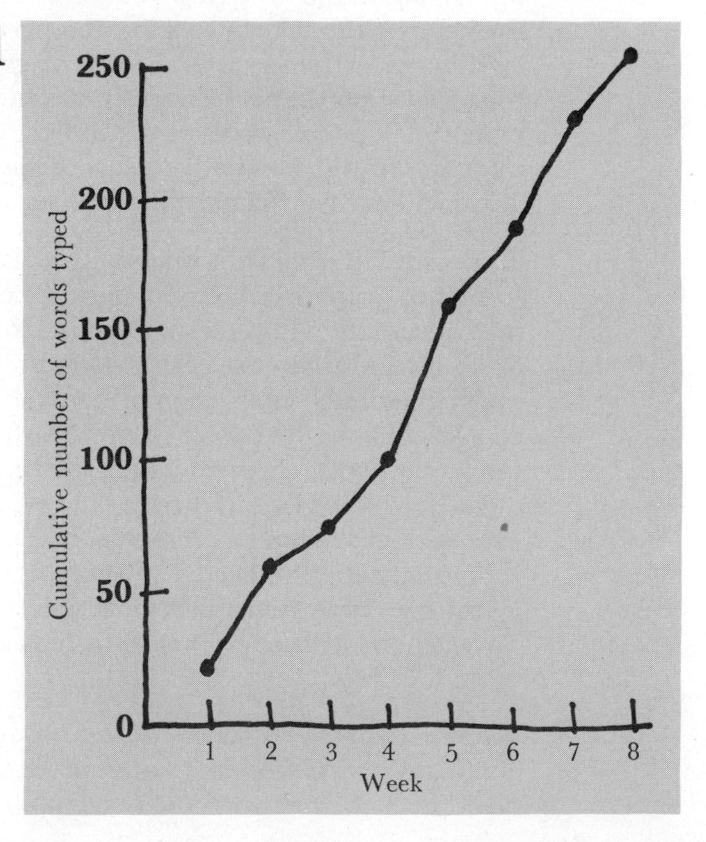

Louie's mistake in reading this chart is understandable. Like most of us, he is accustomed to interpreting a rising line on a graph as good and salubrious and a falling line as unfavorable, at best, and catastrophic at worst. There's only one thing wrong with interpreting cumulative curves in this way. *As long as there are any gains whatever* (number of words typed in Margaret's case) *the line must go up. It cannot go down.* At the very worst, it will parallel the horizontal axis when her typing output drops to zero during any given period.

But there is an infinitely more subtle type of interpretive error that cumulative graphs encourage. Since they appear rather smooth and climb steadily from left to right, they seem to reflect

fig. 5.12

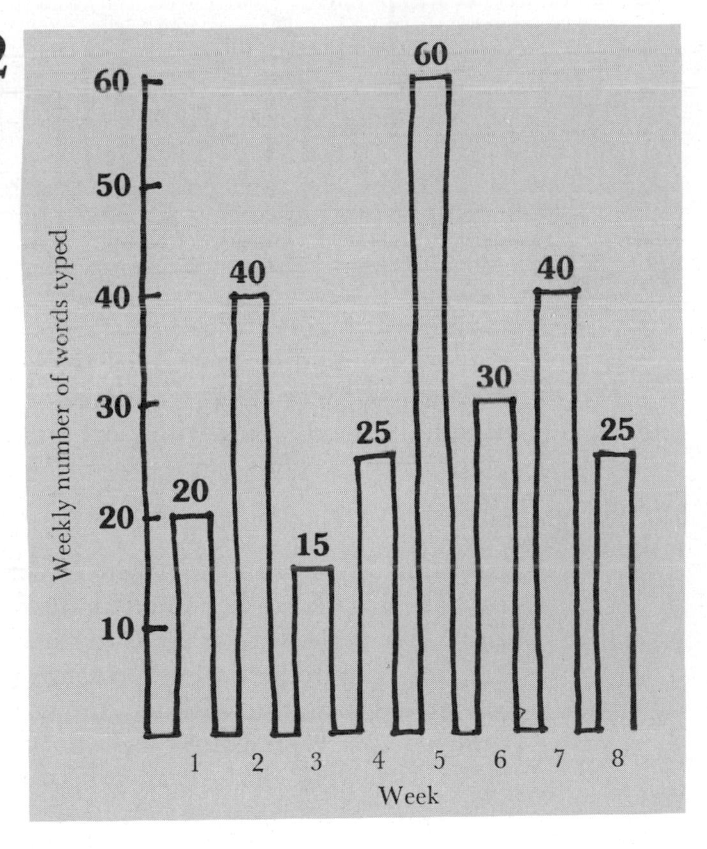

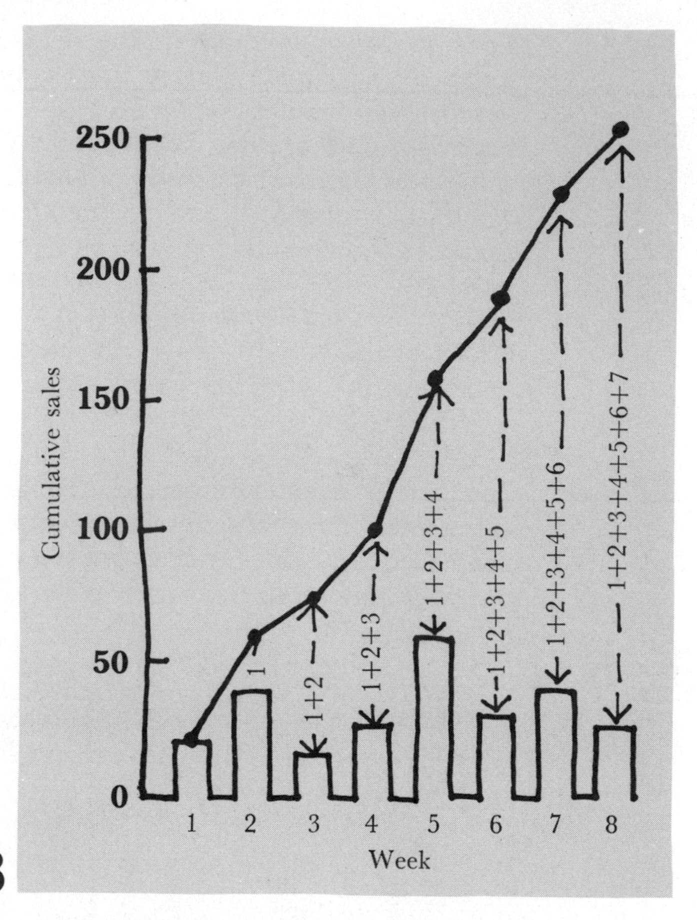

fig. 5.13

stable performance. When a fellow state senator in an adjoining district perused the graph, he commented, "Jeez. That gal is a steady performer all right. A good dependable performer. I could use one hundred like her for my committee work."

Unfortunately, it is in the nature of cumulative curves to appear stable. The cumulative adding of all prior performance tends to smooth out even large differences in performance from one time period to another. In fact, the cumulative curve we have been examining was based on the week-to-week performance figures shown in Fig. 5.12.

Incidentally, in the event that you encounter a cumulative curve and would like to derive the original time-by-time performance for it, the procedures are very simple (see Fig. 5.13).

To derive the original time-by-time performances from a cumulative curve:

1. Measure the height of the curve for the first time period. Subtract this amount from the curve representing the second time period. The difference is the performance during the second time period.

2. Measure the height from the horizontal axis to the rising curve for the second time period. Subtract this amount from the curve representing the third time period. The difference is the performance during the third time period.

3. Repeat these procedures until the original performance at each time period has been derived.

Chapter Six: Ratios, Proportions, Percentages, Index Numbers, and Hocus Pocus

Sometimes the data we collect consist of quantities—heights, weight, speed, temperature, and the like. (Some of the descriptive techniques used with so-called quantitative variables are discussed in Chapters 7 through 9.) At other times we collect data on qualitative variables—sex (male or female), car manufacturer, color, candidates for political office, or yes/no attitudes on a social issue. Such data consist of head counts, e.g., the number of people purchasing Fords versus Chevrolets versus VWs; the number of straw votes for candidate A versus candidate B versus candidate C. The descriptive statistics used with head count data are ratios, proportions, and percentages. As we shall see, these measures are useful ways of summarizing data. However, used improperly they provide abundant opportunities to commit statistical mischief.

The Civilized Head Hunters

One of the favorite research procedures of the behavioral and social sciences, marketing organizations, advertising agencies, and political parties involves the head-count, or survey, technique. We are all familiar with the Harris and Gallup pollsters who ask carefully selected respondents all manner of different questions, e.g., "If you were voting right now, which of the following candidates would you favor for the office of president of the United States?" By dividing the total number of respondents into the number choosing each candidate, we arrive at a most useful statistic—the proportion favoring each candidate. If we are more comfortable dealing with percentages, we multiply the proportion by 100.

Let's look at an illustrative example. In Centertown, U.S.A., a carefully selected sample of 1000 registered voters is asked the question, "Which of the following four candidates for the office of the mayor would you vote for if the election were held right now?" The hypothetical results are shown in Table 6.1.

table 6.1 Hypothetical head-count of 1000 respondents in Centerville, U.S.A., to the question, "Which of the following four candidates for the office of mayor would you vote for if the election were held right now?"

Candidate	Number favoring	Proportion favoring	Percentage in favor
A	186	0.186	18.6
B	208	0.208	20.8
C	435	0.435	43.5
D	74	0.074	7.4
Undecided	97	0.097	9.7

For obvious reasons, the results of such polls are extremely useful to politicians and to political parties. If their main concern is winning the election, they want to put your campaign money where the strength is. On the basis of Table 6.1, candidate C appears to be strongest in the race.

However, if funds allow, polls of this sort can be conducted on a regular basis. Changes in preference from survey to survey can be used to assess the progress that each candidate is making or to evaluate the effects of a change in campaign strategy. For example, let us assume that candidate C agreed to participate in a series of television debates with candidate B. Following the debates, another survey is conducted among the city's voters. The results of this hypothetical poll are shown in Table 6.2.

table 6.2 Hypothetical poll showing number and percent favoring four political candidates, previous percent favoring each, and change in percent favoring each candidate.

Candidate	Number favoring	Percent favoring	Previous percent favoring	Change
A	75	7.5	18.6	−11.1
B	336	33.6	20.8	+12.8
C	440	44.0	43.5	+ 0.5
D	129	12.9	7.4	+ 5.5
Undecided	20	2.0	9.7	− 7.7

It is readily apparent that the fortunes of candidate B have shown a dramatic rise following the television debates. An interesting fact is that they appear to be at the expense of candidate A and the undecided group rather than through losses incurred by the hapless candidate who agreed to the head-to-head confrontation. You might well imagine that candidate C's campaign manager would be thrown into a state of panic by these results, perhaps leading to the cancellation of future televised debates.

Be that as it may, the percentages shown in Table 6.1 represent, in the jargon of statistics, a *distribution ratio*. This type of ratio gets its name from the fact that it shows how the percentages are apportioned among the various classes or categories. In our example, each candidate constitutes a class. One of the nice features of a distribution ratio is that it is very difficult to engage in statistical monkey business with it, since the combined percentage is always 100; there are no such esoteric percentages as 200%, 500%, or 1000%, which can be obtained with some of the other types of ratios that we will be discussing soon. There are really only two types of skullduggery that the unscrupulous can use with distribution ratios in order to befuddle the general public: An unrepresentative or biased sample can be selected and the number of observations can be so small that the percentages are unstable at best, and meaningless at worst. Both are beautifully illustrated in a comic strip whose passing I deeply regret. I am referring to Pogo and his cast of characters in the Okefenokee Swamp. One of the personalities decided to run for the presidency of the United States. Another (I think it was Owl) conducted a poll in order to assess his chances. The results were astounding—a landslide victory for the Okefenokee candidate! Little was made of the fact that only the immediate family of the presidential aspirant was polled (some not of voting age). But Pogo reasoned something to this effect: "Nevertheless, five out of five is 100 percent." I might add, so is one out of one.

But if you really want to have fun with ratios and their various derivatives, look to indexes that express change in terms of percentage of gain.

Making a Loser Look Like a Winner

Take another look at Tables 6.1 and 6.2, concentrating your attention on candidates B and D. Note that between the first and second poll candidate B gained 128 votes; D gained 55. B's gain in numbers was more than twice as great as D's. Candidate B improved by almost 13 percentage points between polls; D's improvement (5.5%) was less than one-half as great.

Conclusion: In relation to D, Candidate B fared well. Right? Wrong, if you listen to D's campaign manager. He pretended to be ecstatic when the results of the second survey were announced. Let's eavesdrop on a press conference he held immediately afterward.

Reporter: "I understand you are pleased with the results of the mayoral poll."

D's manager: "Why, of course, who wouldn't be? My boy is really on the move now. Of all the candidates, his popularity is increasing more rapidly. In this respect, we're number one, nombre un, numero uno."

Reporter: "I don't understand how you can make this claim. B obtained 2.6 times as many straw votes as D, his gain between the first and second survey was 2.3 times as great as D's, and he showed the biggest gain of all candidates in the change column."

D's manager: "Friend, you have known me for a long time. Have I ever been anything but forthright with you?"

Reporter: "Well . . ."

D's manager: "Let's look at it this way. B received 208 straw votes in the first poll and 336 on the second. Right?"

Reporter: "That's what I just finished saying."

D's manager: "Bear with me. B's gain was 128 votes. Now let's express this gain as a percentage of gain. In other words we'll divide the gain by the number of votes B obtained on the first survey, and multiply by 100. That comes . . . let's see . . . to a percentage gain of 61.54. Impressive, even I must admit. B's going to be a tough nut to crack, but my boy is running faster right now than anybody else."

Reporter: "I still don't figure . . ."

D's manager: "Now let's calculate the percentage of gain for my boy. On the second survey he obtained 129 votes, contrasted with 74 on the first. Right? His gain was 55 votes. But his percentage of gain, that's another story. Divide 55 by 74 and multiply by 100 and what do you get?"

Reporter: "I'm not much at math."

D's manager: "74 percent! That's what you get. Right now my boy is running about 12.5 percent faster than B! You still wonder why I am jubilant over the survey results?"

All of this proves that, if you start low enough, any gain can be made to look like a big gain. Back in the days when growth was considered a national asset, almost synonymous with a natural resource, I drove through several communities that claimed to be the most rapidly growing in the United States. It's an easy job if you're small enough to begin with. It would be hard for a city like New York, for example, to match the growth rate of a town that jumped from a population of 10 to 15 in a year. To match this growth rate of 50 percent, New York City would have to add about 4,000,000 people to its teeming streets in a single year.

But I'm just getting warmed up. There is a type of ratio, called an index number or time ratio, that is extremely popular with economists and people in business. When used properly, it is an enormously useful way of measuring change (growth or decay) over time. Many of our major economic indexes—the consumer price index (CPI), the wholesale price index, the gross national product (GNP), and industrial production, to name a few—are commonly expressed as time ratios. In case you didn't notice, the Double Whammy Graph (Fig. 5.10) was based on two popular time ratios, CPI and the purchasing power of the dollar.

But, oh, can you do a lot of mischief if you start monkeying around with the base. Let's take a closer look at index numbers.

At a one hundred percent rate of increase over the past year, we are the most rapidly developing community in the nation.

Time Ratios: You're Decaying Faster than I Am

If you want to express the change in a series of values that are arranged in a time sequence, the time ratio is the statistic of choice. In a *fixed base time ratio,* which we will be looking at here, you select a given time period as your base year. Then the values for that year are divided into the values of each other year and multiplied by 100 in order to express the change as a percentage.

To illustrate, Table 6.3 shows the cost per pound of imported canned beef for selected years since 1950. The third column shows the time ratio, with 1950 used as the base year. Rapid inspection of the time ratio reveals that the cost of imported beef more than doubled between 1950 and 1972.

But who's to select the base year? Aye, there's the rub. The particular year that you choose to represent your base year can have a profound effect on the appearance of all of your subsequent figures. For example, if you want to show a

table 6.3 Time ratio for the cost of imported canned beef for selected years between 1950 and 1972. (*Source: The American Almanac for 1974, The Statistical Abstract of the U.S.* New York: Grosset & Dunlap, 1973.)

Year	Cost per pound	Time ratio
1950	.311	100
1955	.321	103
1960	.328	105
1965	.357	115
1970	.425	137
1972	.632	203

period of very rapid growth in some particular industry, you would pick as your index year one in which the performance of that industry was at its very lowest. If, on the other hand, you wanted to give just the opposite impression, that the industry was not doing well, you would purposely select the year in which the industry had its best performance. Abracadabra, you would now convey—with the very same data—a distinct impression of relatively poor performance over recent years.

table 6.4 Violent crime rates per 100,000 population for selected years between 1960–1974, and time ratios using two different years as the fixed base. (*Source: The American Almanac for 1976, The Statistical Abstract of the U.S.* New York: Grosset & Dunlap, 1975.)

Year	Violent crimes, rate/100,000 pop.	Time ratio (Percentage change) Fixed base, year 1960	Time ratio (Percentage change) Fixed base, year 1970
1960	160	100	44
1965	199	124	55
1967	251	157	69
1968	297	186	82
1969	327	204	90
1970	362	226	100
1971	394	246	109
1972	299	187	83
1973	415	259	115
1974	459	287	127

Now let's see how this is done. Shown in Table 6.4 are the violent crime rates per 100,000 population for selected years from 1960 to 1974. If you simply look at the crime rate, it is quite obvious that the rates are increasing rather alarmingly. Now one can produce quite different impressions if one simply selects different years as the fixed base year and calculates percentages of change from that year. If you look at Fig. 6.1, you will see two plots. One plots the time

fig. 6.1

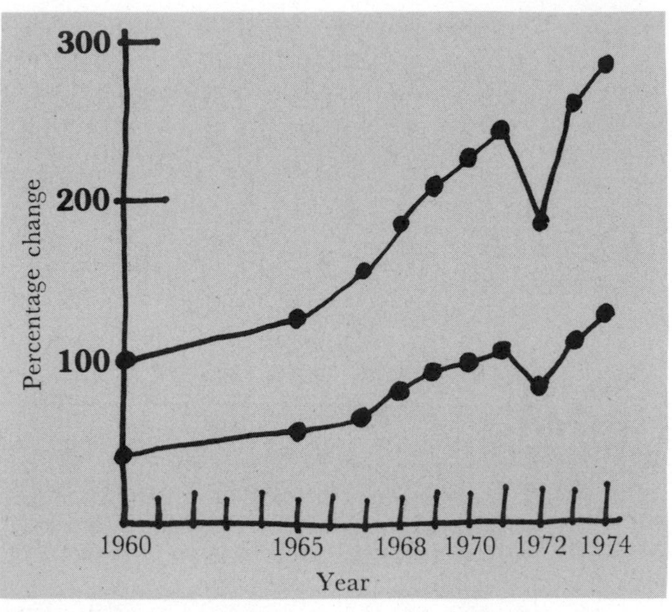

Rate of growth of violent crimes. The impression given by the graph depends on the year selected as the base year. When 1960 is used as the fixed base, the crime rate increase appears explosive. When 1969 is used as the base year, the growth rate appears more modest.

ratio for the violent crime rates for the selected years of 1960 to 1974, and we use 1960 as the base year. In the second we use precisely the same data but we shift to 1970 as the base year. Inspection of the two line drawings clearly shows a huge rate of violent crime increase when 1960 is used as the base year. When 1970 is used as the base year, the rate of increase looks much more modest.

box 6.1

When a 7.2 percent point increase is not a 7.2 percent increase.

Do you remember the Yom Kippur war back in 1973 and the subsequent Arab oil embargo? Do you remember the long gas lines, the short tempers, and the midnight walks to the no-longer-so-friendly gasoline station attendant for a secret refill of the five-gallon gas can? Do you also recall the lugubrious protestations of the oil companies that they too were the innocent victims as they reeled under the merciless blows of unprecedented profits? While the sharp edge of these memories has begun to fade, the aftermath of the petroleum price increases has continued unabated. In our dealings with the beloved local utility, we have become accustomed to the fuel price adjustment, the energy conservation adjustment, and the almost continuous requests for rate increases "to match the general inflationary spiral." The consumer price index has become the favorite instrument for a whole host of pricing and wage decisions by every conceivable organization and its grandmother. However, confusion in the understanding of this index could lead to a massive rip-off of the hapless consumer. This is how the scenario goes.

Imagine that the Bureau of Labor Statistics has just released the bad news concerning the increase in the consumer price index for the previous month. Your local ____________ (you fill in the name of your favorite villain) requests a rate increase to match the hike in the CPI. The spokesman for ____________ reasons as follows: "The CPI went up by six-tenths of one percent last month. That equals an annual rate of increase of $12 \times 0.6 = 7.2$ *percent. We are, therefore, requesting an annual rate increase of 7.2 percent."*

Sounds reasonable enough, if rate increases are ever reasonable. However, our friend is confusing the number of percent points *of increase with the* percentage *increase. But let's not forget that the CPI used 1967 as the base period. Only during that year was the CPI at 100. Only during that year would an increase of 7.2 percent points equal a percentage increase of 7.2. The CPI has been galloping upward ever since.*

Let's say that, at the beginning of the previous month, the CPI had been at 150. During the month, it increased to 150.6. This is an increase of six-tenths of one percent point, or an annual rate of increase of 7.2 percent points. So far so good. But to find the percentage increase, we must use 150 as the base. The annual rate of increase is $(7.2/150) \times 100 = 4.8\%$*. Our friend should really have requested this much of a rate increase.*

An honest mistake? Maybe. But why do honest mistakes so often wind up getting the consumer in the neck?

I am waiting anxiously for this year's profit statement from the Seven Sisters. Oil company profits immediately following the oil embargo in 1974 were so enormous that I would be shocked out of my mind if they failed to use 1974 as their base year. With such a base, they should now appear to be on the verge of beggary.

But there are other ways to manipulate the base to one's advantage. One of my favorites is what I call "stealing by the base."

Stealing by the Base

Any baseball enthusiast will understand immediately what is meant when the announcer proclaims, "Joe Morgan just stole second base." What the baseball fan may not realize is that there are many ways, besides playing baseball, of committing larceny with bases. I just happened to have in my collection of tapes a conversation between the redoubtable Sam Stickfinger and his financial consultant, Esab Rewol. Esab explains how a markup of 100% can become a 200% reduction during a sale in which the price of an item is cut 25%. All that is required is a little flexibility in selecting the base from which the calculations are made.

Sam: "We pay 50 cents for the item and sell it for $1.00. I thought of advertising that our markup is only 100%."

Esab: "That's suicide, Sam. Nobody will buy an item when they know the markup is 100%. But your markup is really only 50%."

Sam: "How do you figure that?"

Esab: "The item costs the customer $1.00, right?"

Sam: "Right."

Esab: "The profit is 50 cents, right?"

Sam: "Right."

Esab: "Well, 50 over 100 is 50 percent. It's just a matter of which base you use, 50 or 100. But you never really sell the item for $1.00, do you?"

Sam: "Oh no, it's always on sale. We take 25 cents off and advertise a 25% reduction in price."

Esab: "Bad, Sam, bad. If you take off 25 cents, the new price is 75 cents. Right?"

Sam: "Right."

Esab: "Well, 25 cents over 75 cents is one-third of the new price. You've got a one-third, or 33% sale."

Sam: "Fantastic! I've been practically giving it away without knowing it."

Esab: "It's worse than you think, Sam. The cost to you is 50 cents, right? Nobody can expect you to stay in business if you ever went below that price, right?"

Sam: "Right."

Esab: "So we should really be advertising in terms of reduction in profits. See?"

Sam: "I'm beginning to see. At $1.00 our profit would be 50 cents per unit. We have reduced our profit by 25/50 or 50 percent. It's really a 50% off sale!"

Esab: "Yes, if you want to be completely honest about it . . ."

Sam: "Of course I want to be honest. But out of idle curiosity . . ."

Esab: "Well, if you don't mind bending the truth ever so slightly . . ."

Sam: "Bending slightly is honest. Bending big is dishonest."

Esab: "Well, your profit is 25 cents. Your reduction is 25 cents. If you take the reduction and divide by the profit . . ."

Sam: "You get a 100% reduction. Why those damned customers, ripping me off like that."

Esab: "Yes, we'll call it a 'Rip Off Sam Sale'. Oh, yes, by the way, have you ever considered putting a price tag of $1.25 on that item?"

Sam: "No. Should I?"

Esab: "Well, when you have your sale and reduce the price to 75 cents, you can claim . . . let's see now . . . 50/25 × 100 . . . that's a 200% reduction!"

Sam: "And Joe Snopes down the block can only advertise that he sells below cost."

Esab: "You can do better than that. The usual cost to the customer is $1.25, right?"

Sam: "With our new price tag, that's right."

Esab: "Well you're going to reduce it to 75 cents, 50 cents below cost. That's 50/125 × 100 = 40%. Just think of it, 40% below cost!"

Sam: "I like 200% reduction better."

Esab: "Why?"

Sam: "It sounds like I'm giving it away, which I am."

Well, have you had enough? Let me tell you just one more. In my opinion, the next one, which I call "Paradoxical Percentages," is the pièce de résistance on the menu of statistical deception.

Paradoxical Percentages

Let me pose a problem that is routine in various state and federal Human Rights Commissions. Suppose you suspect that a given business, industry, municipality, or what have you is discriminating against women in its hiring practices. How would you go about investigating the charges?

That's right. You would investigate the number of qualified men and women who have applied for various positions. If you found a greater percentage of the qualified males than females hired for available jobs, you would have pretty good presumptive evidence of discriminatory practices, right? I'm going to surprise you by answering my own question. My answer is a firm, categorical "not necessarily." How come?

Let me show you some hypothetical data in which the proportion of females accepted in each job category is actually higher than the proportion of males accepted in each of these categories. Nevertheless, when the proportions of males and females accepted for the various positions are combined over all of the categories, it is found, as if by magic, that the females have a much lower overall proportion of acceptances. So if you were to look at the overall figures you'd be led inexorably to the conclusion that there is discrimination against females. However, if you looked at the same evidence category by category you'd be led to precisely the opposite conclusion, namely, that there is discrimination against males. Let's look at these data in Table 6.5.

table 6.5 Hypothetical data showing the number of male and female applicants for various teaching categories, the number accepted, and the percent accepted.

Job category Teacher	Male No. Appl.	No. Accp.	% Accp.	Female No. Appl.	No. Accp.	% Accp.
Grades 13–14 (2-yr coll)	150	30	20	40	16	40
Grades 11–12	200	70	35	50	35	70
Grades 9–10	100	15	15	50	15	30
Grades K–8	50	2	4	600	48	8
Total	500	117	23	740	114	15

Take a careful look at the table. It has been purposely constructed so that in every single category, the percentage of women accepted is double the percentage of men accepted. Nevertheless, when all of the applicants among the males are combined, and the number is divided into the number of males accepted, and the same thing is done with the female category, you find that in fact the overall percentage of males accepted (23%) is higher than the overall percentage of females accepted (15%). How is it possible to obtain this sort of numerical sleight-of-hand? Very simple. If you look at the table again you will find that most of the female applications, 600, were for positions in Grades K through 8. Conversely, very few of the male applicants were for those positions. It turns out that the number of openings for grades K through 8 is the smallest. What you have then is a situation in which an overwhelming majority of the women were applying for "difficult-to-get" positions. On the other hand, most of the men

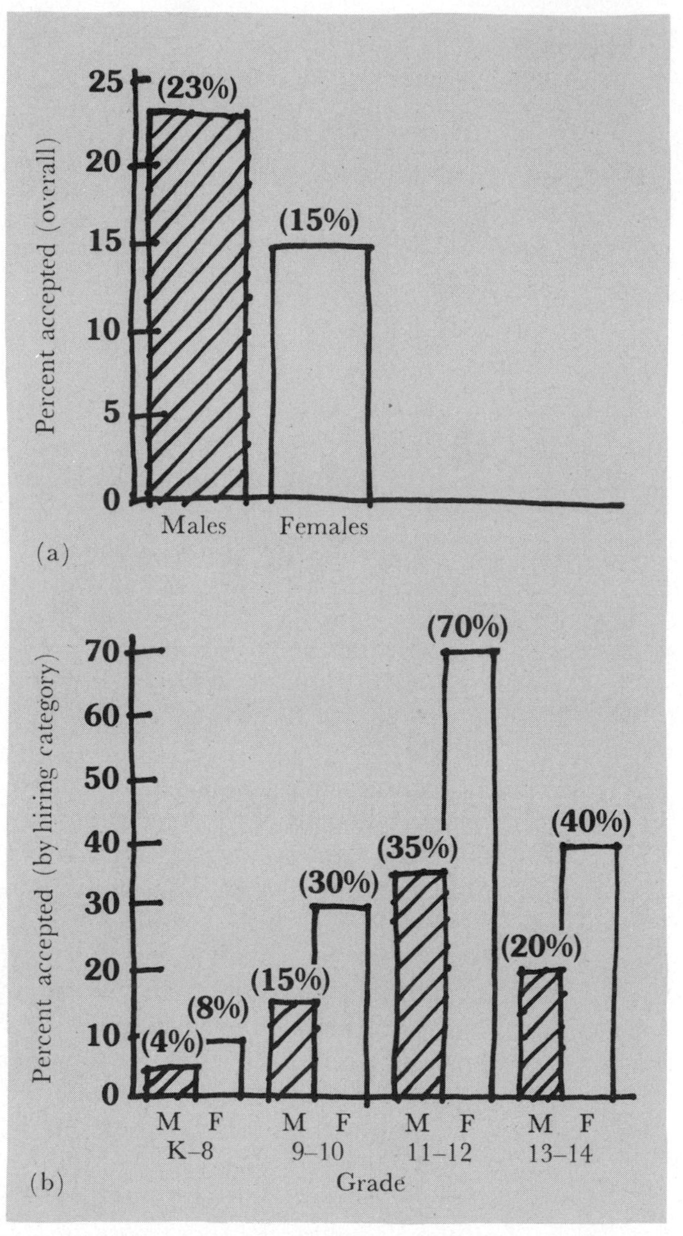

fig. 6.2

Paradoxical percentages. (a) When we calculate percentages of equally qualified males and females accepted for teaching positions, we find substantial evidence of discrimination against females. (b) However, when the results are broken down into four separate hiring categories, overwhelming evidence of discrimination against males is found. See the text for a more complete explanation.

were applying for positions that were more plentiful. When you combine all of these categories in which there are different numbers of applicants for different positions, the low percentage of hirings of females in the K–8 group is devastating to the female teachers. Why? Better than 80% of the female teachers applied for the very positions where the number of openings was extremely low in relation to the number of applicants.

Now take a look at Fig. 6.2. It tells the whole story much better than I can do it with words.

Please don't consider me a "male chauvinist pig" for using this example. I assure you, I'm not a chauvinist.

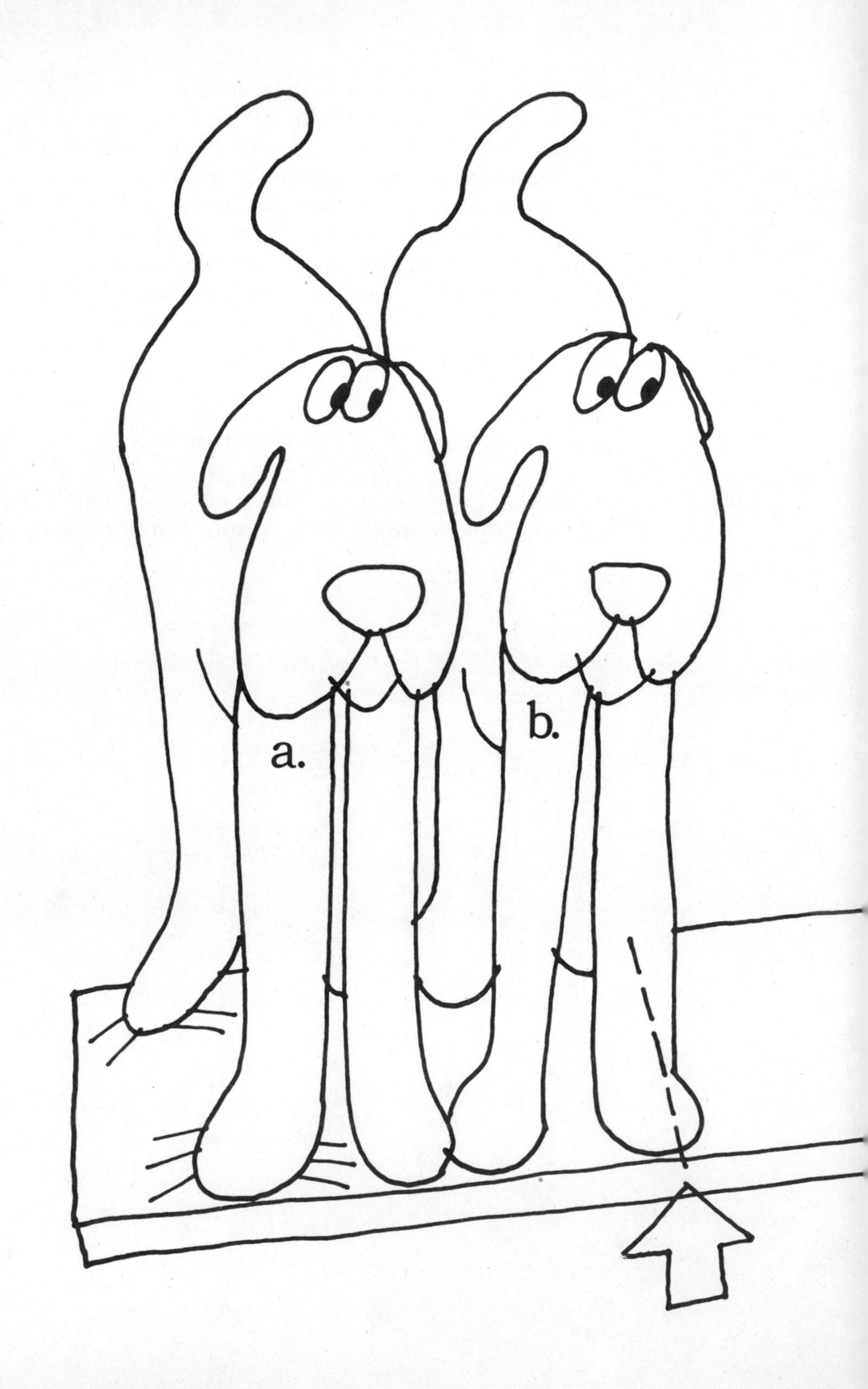
a.
b.

Chapter Seven: What Does the Mean Mean?

How good is your imagination? Can you imagine yourself as the head of a local labor union making preparations for a strike? You can? Good. Then imagine me as the well-fed and somewhat corpulent corporation executive preparing to do battle with you over the latest outrageous union demands. And, of course, you are the gaunt, emaciated representative of the downtrodden laboring class, complete with two days of chin stubble, cavernous eyes, and hollow cheeks. Of course, we know this image is a fiction. We're both pretty portly these days. In fact, as head of the labor union you probably earn more than I do when fringe benefits are considered. And certainly your tennis club is equal to mine. But don't you sometimes long for the good old days when you were a Bolshevik

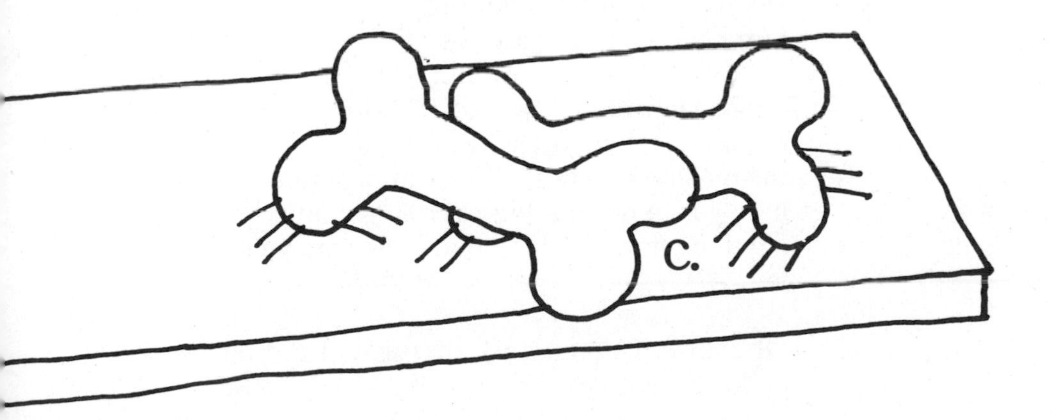

agitator—a bad guy—and I was the knight in shining armor trying to protect our system from incursions of the Red tide? We'd call out the goon squads and you'd plant a few bombs. A few heads would get bashed in the process but we'd work something out somehow. But now it's lawyers and more lawyers—countless hours of lawyer gobbledegook. Ah, the fun is gone. I sometimes think labor unions were invented by lawyers, just so they could collect a fee. I mean they invented everything else—possession of private property, marriage, divorce, cars, babies, accidents—so why not labor unions?

But I am digressing. The point I am trying to make is that the armamentarium in the union–management battles has changed radically over the years. A big part of the skirmishing now takes place between the public relations officers of both sides. And what is their main weapon? You guessed it. The Statistic. Not necessarily the statistical lie but merely the statistic deployed in such a way that the same effect is achieved.

Let's see how this works. You, the big labor boss, call on your public relations man. You say to him, "Get together some statistical proof that our laborers are underpaid. I've already reserved a full-page ad in *The Times* and I've written the ad. Now all I need is the numbers to fill in the missing spaces." (Boy, I'm beginning to hate you. You're a big overblown oaf who thinks anything goes. Well, if that's the way you want it . . .)

Of course, I don't work that way. I am honest and I always seek the truth. I call my public relations man and ask him to dig up the data on salaries, fringe benefits, and all that stuff: "Most of all, I want to know what the average salary is around here. I'm sure it's one of the highest in the industry and I think we should be telling it to everybody but the stockholders."

The answer comes back promptly with the pure tones of Berlioz' "Ranz des Vaches" (I like to show off a little bit) : "Our average hourly salary, exclusive of fringe benefits, is $5.50."

"Ho, ho," I say, "that will cook their little goose-flesh behinds. Here we are fighting spiraling inflation, sacrificing, cutting our profit gains to under fifty percent of last year, and those crumbs are looking for raises. It's unpatriotic, undemocratic, and anticapitalistic."

Imagine my surprise when I open *The Times* on the following morning and see the union ad, "Peons at Wankee receive lowest wages in the industry. At $3.00 an hour, many qualify for welfare."

Now, my friend, let's step out of our adversarial roles for a moment and examine this situation. Management says that the average hourly wage is $5.50, labor claims it is $3.00. Is labor lying, management lying, or are they both lying?

The strange thing is that both may be telling the truth. In fact, they would be crazy not to tell the truth when the term *average* provides a loophole much bigger than the eye of a needle—we could easily put a camel through this one!

The term average is one of those ambiguous words meant to tell you something about the middle of a distribution of numbers, of scores, of wages, etc. The only trouble is that there are three different measures that describe the center of the distribution. They are called *measures of central tendency* and they are defined in different ways. With many data, this difference in definition will not make a difference. With others, such as wages, the difference is as great as the disparity between gross income (several million dollars) and net adjusted income (zero) on a rich man's tax return.

Now, let's take a peek at the hourly wages of employees at Wankee. To simplify matters, we'll show only a small representative sample of the total. As a matter of actual fact, Wankee employs thousands of laborers in its East Coast plant. A large number are unskilled and obtain the lowest wages. A select few are highly skilled, possess advanced degrees, and command substantial hourly wages.

Name of employee	Hourly wage	
110 15 2436	2.50	
109 16 4134	3.00	Mode ($3.00)
015 62 3343	3.00	
101 45 1362	3.00	
515 60 4142	3.00	
612 45 3627	4.00	← Median ($3.50)
413 21 6561	4.00	
218 35 4425	4.50	
806 56 7132	5.00	← Mean ($5.50)
Mr. Parsons	23.00	
Total	55.00	

Wage and salary figures characteristically show much greater extremes at the high end than at the low end. After all, you can't earn less than zero income but there is no limit at the other end, particularly if you're Lamar Hunt or a Persian Gulf oil czar.

The three measures of central tendency are the mean, the median, and the mode. We are all familiar with the mean. When we were in school and had to calculate the arithmetic average of a set of grades, we added them together and then divided by the number of grades. So, if we got 90, 70, 80, and 80, the sum would be 320. Divided by 4, this yields a mean of 80. In our example with Wankee, the sum of the hourly wage of the ten employees is $55. When divided by ten, a mean of $5.50 is obtained. That's where I got my "average" when I played the role of an executive of Wankee. Note one important characteristic of the mean: Every score enters into its determination. If there are extreme scores at one end of the distribution (in our example, Mr. Parsons' salary), the mean is pulled in the direction of those scores. In this rather exaggerated case, the mean is a deceptive "average" —it represents no one—and I deserve to be chastized for this bit of statistical deception.

But don't think that lets you off the hook. No siree/ma'am (cross out the one that doesn't apply), you're just as big a crook as I am. You chose the mode, which is the score that occurs with the greatest frequency. You're taking advantage of the fact that Wankee employs mainly unskilled workers, all of whom earn wages that are consonant with their contribution, in other words, low on both counts. Your "average", my friend (and I use the word advisedly), represents only our lowest paid employees.

If we were both dedicated to honesty, we would select the median as our measure of central tendency in this case. Why? The median is the middle score. You find it merely by counting down until you locate the score that is exactly in the middle. Half of the scores are above it and half are below. In the present example, the middle score doesn't really exist so we arbitrarily select a value that lies halfway between the fifth and sixth scores. This gives us a median hourly wage of $3.50. Note that the median provides a fairly good representation of the majority of the wage earners. Note also that unlike the mean, the median is unaffected by extreme scores. If Mr. Parsons had earned

$1,000,000 an hour (and revealed his true name to be Sheik Yemani), the median would have remained precisely the same. Generally speaking, the median is the measure of choice when there are extreme scores in a given direction, as with salary and wage figures. We speak of such distributions as being *skewed*. If you think of the word askew—off balance—you've got the right idea.

But the fun and games are just beginning. You've all heard of the double standard and mistakenly thought that it referred to male-female freedoms. The double standard consists of using two different statistics at the same time to make your point. But why am I telling you this? You know all about it. I forgot to tell you we had your phone bugged. In case you're a bit hazy, let me reproduce a bit of one of the taped transcripts to refresh your memory. The conversation is with your West Coast counterpart.

You: Say, Harry, it looks as if we're going to have a bit of a battle with those (expletive deleted) at Wankee over salary negotiations. Could you give me an idea of the average hourly wages out there?

Harry: No sweat. We're getting our (expletive deleted) butts ready for negotiations here so I've got them right at my fingertips. Er . . . let's see. Ah, here they are. The average hourly salary here is $2.90? Why, I'll be a (expletive deleted)!

You: That's all?

Harry: A lot of uneducated labor, you know.

You: Yeah, I forgot. How much is management claiming as the average?

Harry: About $5.30.

You: Would you mind sending me their figures? Forget about your own. It's only fair that we use figures from both sides. My figures for the East coast and management's for the West coast. That way, no one can accuse us of being biased. The figures also have a nice ring to them. Our guys sweat for $3.00 an hour and yours get about $5.30. By God, we're being screwed.

Harry: I'll send you their figures under one condition.

You: O.K.?

Harry: Yes—you send me the figures that your management has compiled. What's good for the goose is good for the gander.

You: (expletives deleted)!

End transcript

Do You Mean We Shouldn't Use the Mean?

Now don't get the wrong idea from our previous example. Most of the time the mean is a pretty good measure of central tendency. It's just that when the distribution is badly skewed, the mean simply does not tell you much about the most typical or representative scores. But even in those cases, the mean provides information not available from the other two measures. For example, what if you wanted to know how much Wankee pays its ten employees each hour and the only information you have is that the mean is \$5.50, the median is \$3.50, and the mode is \$3.00. Here the fact that the mean is based on arithmetic processes—adding, subtracting, multiplying, dividing—makes our problem a lead pipe cinch.

Remember:

$$\text{Mean} = \frac{\text{Sum of scores}}{\text{Number of scores}}.$$

By the simplest of algebraic manipulations, we can find the unknown total:

$$\text{Sum of scores} = \text{Mean} \times \text{Number of scores}$$

$$\text{Sum of scores} = \$5.50 \times 10$$

$$= \$55$$

Here's a beauty that my wife just clipped out of the Sunday Newspaper Magazine (Parade).

MR. AVERAGE The average American male is 45 years old, has a 42-year-old wife and two teen-age children. He eats half a ton of food annually--126 lbs. of bread and butter, 102 lbs. of sugar, 247 lbs. of milk and cream and 287 eggs--and drinks 16 cups of coffee a week.

He can expect to be involved in at least one minor traffic accident every four years and perhaps a bad one every 20 years.

Mr. Average views 11 films a year and spends about 3½ hours a day in front of his color television set.

My wife likes this one. Says she, "Since you're going on fifty one, nobody can accuse me of being married to just an average man. That's a claim I couldn't have made six years ago!" (Source: Parade: *The Sunday Newspaper Magazine, New York: Parade Publication, 4 May 1975, p. 4.)*

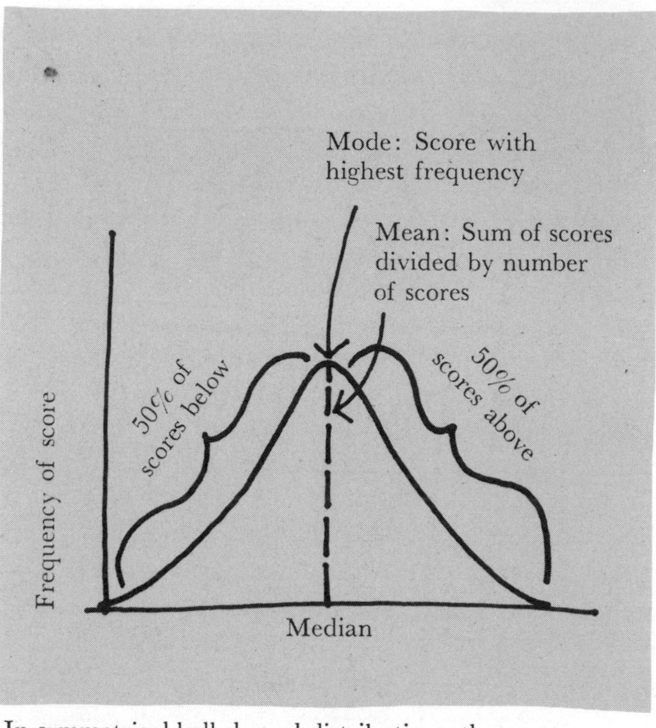

fig. 7.1

In symmetrical bell-shaped distributions, the mean, median, and mode are all the same.

This mathematical characteristic of the mean makes it a very valuable statistic, in both descriptive and inferential statistics. When you add to this the fact that many of the things we measure in this world tend to distribute themselves in a symmetrical bell-shaped fashion, it really doesn't make much difference which measure you use—the mean, the median, or the mode. They are all the same. You can see this more clearly in Fig. 7.1.

While we're on the topic of averages, let me get one of my favorite gripes off my chest. To be most blunt about it, the average person bugs me. That's right. I hate the average person. Before you get me wrong, let me explain what I mean. Somebody starts out with a few statistical facts. The average (mean) I.Q. of the general population is 100. The average (mean) height of males is 5′8″; of females, 5′4″ (I just made those figures up. I forget what they are and it really doesn't make any difference for the point I am making). The next thing you know a subtle semantic transmogrification takes place and we hear, "The average person has an I.Q. of 100"; "The average man is 5′8″ tall"; and so on, *ad absurdum, ad nauseam*. I don't have any idea what sort of creature the average person is, but of one thing I am sure. Having one attribute that is average does not make the individual average. I know a clod who is 8′2″ tall, weighs 400 pounds, giggles, and has six toes on his third foot. But he has an average I.Q. Does this make him average? Oh, how I hate the average person. For that matter, I hate the average anything.

"We are dumbfounded, Madame. Your baby is 3 feet, 7 inches tall!"
"That's all right, just as long as she's average."
DELIVERY ROOM

Chapter Eight: A Standard Deviation Is Not a Sexual Perversion

Take a look at that cartoon again. It is telling an important story. We tend to get terribly hung up on what's average. We forget that few things are average on any given characteristic that we wish to measure. How many people do you know who are of average height, weight, and I.Q., who come from average families in average crime areas, and who own average houses with average-sized mortgages on average-sized plots? The truth of the matter is that the three measures of central tendency are statistical abstractions. Proportionally, very few members of a given population achieve an "average" value on any measure that we wish to examine. I think we can assure ourselves that very few parents have achieved that wondrous and advanced state of planning in which they have exactly the number of children, 2.1, that statistical tables tell us is the average (mean). The truth of the matter is that most measures in the real world are either below average or above average. Failure to recognize this ineluctable fact of statistical life leads to that delightful state of affairs known as a WOW statement. What's a WOW statement?

Do you know that half of the people in the United States are above average in weight? WOW!

Do you know that half of our children are below average in I.Q.? WOW!

Do you know that half of our people are below average in emotional balance? WOW!

Do you know that half of the Japanese are above average in emotional balance? WOW!

Do you know that half of our cars are below average in safety? WOW!

Do you know that half of foreign cars are above average in safety? WOW!

Prepare yourself for the next one. It is heartrending, a tragedy that science must do something to correct.

Do you know that half of the people in our country die before they reach average age? WOW!

And that half the people in other civilized nations die at or above average age? WOW! WOW! WOW!

It's just as we always suspected. American medicine is like a freight train going down hill without brakes (although the costs are going in the opposite direction with jet propulsion). The reason? An above average number of shows on television are about doctors, nurses, hospitals and the shenanigans that take place underneath those white coats. With all of those doctors on TV, how many are left for us?

Failure to recognize the abstract nature of "average" can also lead to a lot of mischief. Many parents, reading the norms compiled by child psychologists, grimly resolve that, by hook or by crook, their children will not be below average on anything. (And I mean *anything.*) Consideration of this sort leads to another class of shockers which I refer to as OY statements.

Do you know that the average child walks with support at 48 weeks of age? Your Johnny is a year old and he is still grounded on his tush. OY!

Do you know that the average child speaks her/his first word at thirteen months of age? Mary is fourteen months old and still makes bubbles when she says "goo." OY!

Do you know that the average child stops messy-messy by thirty months of age? Your Theosopholus is 34 months old and he still stinks to high heaven. Damn brat!

The truth of the matter is that this world of ours is wonderfully varied. Thank the Creator that we are not all average everything or anything. Of course, if we were it would simplify things. Imagine telephoning the local clothing store, and asking, "Send me an average pair of pants or an average dress," or telling your friendly automobile dealer, "Send me an average lemon." Convenient, yes, but what a crushing bore!

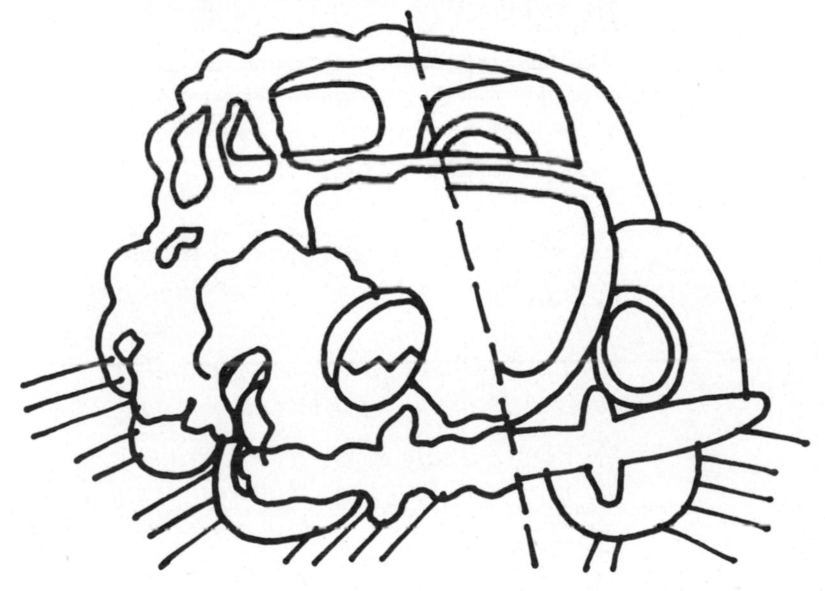

"One-half of foreign cars are above average in safety"

But we are different from one another. Things are different from one another. We are even told that no two snowflakes are ever alike (I'd like to see how anybody would go about proving *that.*) Because of the wide variability of all things measured, it is clear that measures of central tendency have only limited value in describing the totality of events that interest us. Clearly some companion to central tendency—a descriptive measure of the spread of scores about central tendency—is desired. "Why?" you ask.

---

box 8.1 *"Shaving Peak" is Reducing Variability*

At times our preoccupation with averages can cause us to lose sight of the fact that many of the most important workaday decisions are based on considerations of the extremes, rather than on the middle of a distribution. Imagine what life would be like if:

— Our highways were constructed to accommodate the average traffic load of vehicles of average weight.

— Mass transit systems were designed to move only the average number of passengers (i.e., total passengers per day divided by 24 hours) during each hour of the day.

— Bridges, homes, and industrial and commercial buildings were constructed to withstand the average wind or the average earthquake.

— Telephone lines and switchboards were sufficient in number to accommodate only the average number of phone calls per hour.

— Your friendly local electric utility calculated the year-round average electrical demand and constructed facilities to provide only this average demand.

— Emergency services provided average person-

nel and facilities during all hours of the day and all seasons of the year.

— Our man-in-space program provided emergency procedures for only the average type of failure.

Chaos is the word for it. Utter chaos. The fact of the matter is that virtually all of human endeavor must gear itself to meet the extreme conditions known as peak load. If you don't mind my digressing a bit, let me say that the peak load problem is, at once, one of the great challenges and monumental opportunities that we face today. It is because of peak load that many community facilities and services are barely used during certain time periods and are swamped at others. Assuming that the years ahead are sure to place a continued stress on resources, both natural and human, we shall not long be able to sustain the luxury of "gearing up to peak."

The alternative is to raise the valleys and lower the peaks of demand. In statistical terms, the goal is to obtain the same average while reducing the variability. By doing so, we are able to increase and improve our use of existing facilities To illustrate, some years ago my family and I visited Paris to plug the publication of the French edition of The Energy Crisis. *We spent a great deal of our time riding the wonderful Metro (Paris subway system). Besides the quietness of the ride and the attractiveness of the stations (the one at the Louvre actually has faithful reproductions of great works of art and sculpture), we marvelled at the relatively uncrowded condition of the subways during rush hours. A few inquiries elicited an explanation—commerce and industry in Paris operate on a staggered system of opening and closing hours. While Jean-Paul still sleeps, Jacques is on his way to work. The result of this "peak-shaving" strategy is to spread the morning and evening rush more evenly, thereby permitting a more efficient use of rolling stock and sparing many frayed nerves.*

Look for the utilities to solve their present dilemma of perpetual rate increases by finding ways to even out both the diurnal and seasonal demand for electricity.

Imagine that you are a home builder and you know that the median family size is 2.19. (That's what it was in 1972; note that, again, we have the statistical abstraction. I am willing to give considerable odds that there are few families of median size.) You are planning to put in a housing development. How many bedrooms should you provide? Our measure of central tendency does not help very much. What you need is some information about how family size is distributed, spread out, or dispersed throughout the population. What proportions of families are two-person, three-person, four-person, five-person, and so on? Data such as the following (which I took from the *American Almanac of 1976: The Statistical Abstract of the U.S.*) are far more useful.

Size of family	Number in thousands	Percent
2 persons	20,592	37.4
3 persons	11,673	21.2
4 persons	10,789	19.6
5 persons	6,386	11.6
6 persons	3,021	5.5
7 or more persons	2,593	4.7

This information tells us that, were we to build all housing units to accommodate two- and three-person families, we would be neglecting about 41 percent of the population. That's a lot of people and a lot of houses.

Or take another example. Your little twelve-year-old Penelope has just returned from school.

You: How was school today?

Penelope: You know, fine. You know.

You: What did you do?

Penelope: Nothing, you know.

You: Nothing?

Penelope: You know, nothing much. Took a test, you know. You know, one of those tests of scholastic achievement.

You: A test of achievement? Well, how did you do?

Penelope: You know, OK, I guess. You know.

You: You guess?

Penelope: Yeah, you know. I guess so. I got, you know, 55 right. You know.

Now, at this point, do you congratulate Penelope for doing so well, mildly chastise her for not doing better, or excoriate her for doing so poorly? (Or put her over your knee for all of those ridiculous "you knows"!) None of these, of course. You simply have no standard by which to judge a score of 55 correct. So you probe further.

You: What was the class average?

Penelope: Dear parent, you should know that the word average is imprecise and ambiguous (snot-nosed kid!). There are three measures of central tendency—you know—the mean, median, and mode. Which would you like to know?

Now is your turn to show off your erudition.

You: Is the distribution of scores symmetrical or skewed?

Penelope: It's a, you know, standard educational test. You know, that means it's symmetrical. You know. It's, you know, bell-shaped.

You: Well, if it's bell-shaped, what difference does it make which measure of central tendency you tell me? They're all the same.

Penelope: I didn't, you know, know you knew. The mean for the test is fifty.

What do you know at this point? Not really very much. You know that Penny scored higher than the mean. Since the mean and median are identical, you also know that more than 50 percent of the people taking the test scored lower than your daughter. So the question, "How good is a score of 55?", must await additional information about how the scores are dispersed around the mean. So you continue the grilling.

You: What is the highest score?

Penelope: 75.

You: And the lowest score?

Penelope: You know, 25.

Now, you know, we know the range of all possible scores: $75 - 25 = 50$. This tells us something about the dispersion of scores, although not very much. We know that Penelope's score, albeit above the mean, is not near the top. But it's closer to the top than if the range had been 0–100. On the other hand, if the range had been 45–55 . . . oh, blessed nirvana! All of this is made visual in Fig. 8.1.

The range is not usually expressed in terms of the highest and lowest *possible* scores or values, but rather in terms of the highest and lowest values *actually achieved* by a group in which we are interested. It is entirely possible that, in

Penny's class, the highest obtained score was below 75; perhaps even as low as 55. If this were the case, Penny's score of 55 would begin to take on a different appearance.

The value of the range is further complicated by the fact that many of the dimensions that pique our curiosity contain bottom scores but the upper end is wide open. A few very extreme cases will distort our perception of the range within which most values fall.

When we know only the mean

fig. 8.1

When we know the mean and ranges

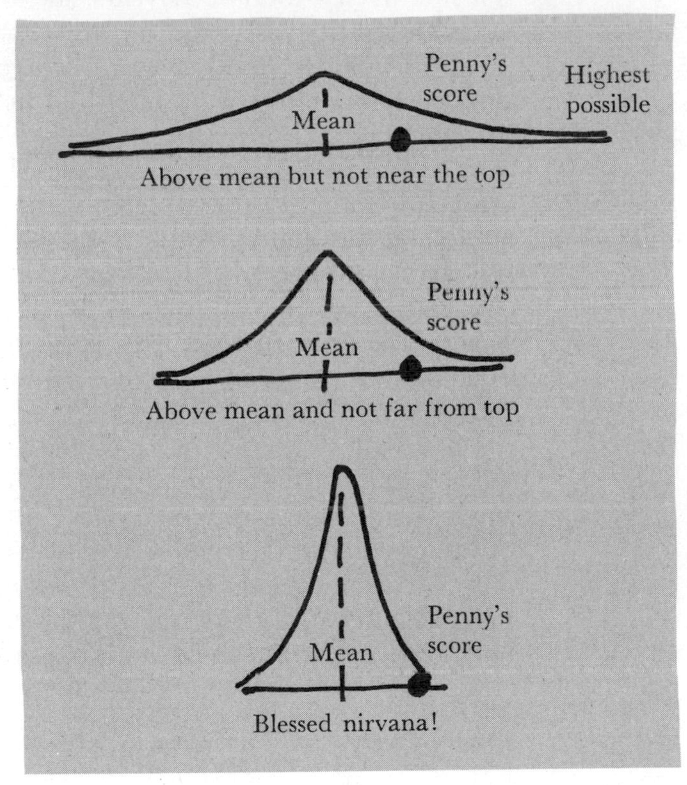

Consider the following.

Income. When this drops down to zero, it's at rock bottom. You can't go lower. But what's the top? Nowadays, I guess the answer depends on how much oil you're sitting on and how fast you can squeeze it out. Certainly the energy crisis is bound to create more than a few billionaires in scattered parts of the globe. It will also leave in its wake many more than a few paupers.

Family size. The number of children in a family cannot fall below zero, but the upper end is far less precise and limited. The *Guinness Book of World Records* reports a substantiated case in which a Russian woman, Mrs. Fyoder Vassilet, gave birth to 69 children. Excuse me for saying this, but I can't resist the temptation. Since Mrs. Vassilet lived a total of 57 years, she averaged 1.2 children per year from the nursery to beyond Medicaid. Stated another way, she averaged one-tenth of a child every month of her life.

These are just a couple of examples. I could cite many more. While athletes of the world are stretching for that 3:50 mile,* the twelve-minute mile still eludes my plodding legs. I am also getting rather portly with increased age, but thank heaven I'll never threaten the world record of 1069 pounds. No piano-case burials for me, thank you.

What's the point of all this? Simply that the range, while useful in depicting the total spread of values on any given characteristic, represents only the two most extreme values. Knowing a measure of central tendency and the range does not allow us to grasp the dynamics of the distri-

* Shortly after this was written, the 3:50 mile was shattered by John Walker of New Zealand at 3:49.4.

bution of scores between the two extremes. After all, knowing the heaviest human being weighed over half a ton and the lightest adult balanced the scale at the weight of a good-sized trout (4.7 pounds) does not tell us an awful lot about the range between which *most* human weights fall. The range is simply the victim of the two most extreme values.

There is a measure of dispersion or variability that is absolutely ideal so long as one condition is met—the scores must be distributed in a bell-shaped fashion, known as the *normal distribution.* You will recall that this is the type of distribution in which all three measures of central tendency are identical. The ideal measure of dispersion or spread under these circumstances is the *standard deviation.* It is ideal for many reasons, but one stands out above all the others: Once you know the mean and standard deviation of a distribution of scores, you know all that is necessary to interpret *any* scores and you can reconstruct the entire original distribution of scores, if you like. Let's examine that statement a little more closely.

Imagine that we have just established radio contact with a civilization out there in space somewhere. (This could well happen within the lifetime of most of us. With billions of stars in billions of galaxies, it seems inconceivable that earth is alone endowed with intelligent life. Indeed, it is probable that there are some civilizations out there that are as advanced over us, in knowledge and technology, as we are over the Neanderthals. If this is so, it is unquestionable that some have been sending out radio signals—perhaps for thousands of years—in hopes of intercepting life elsewhere. What has changed for us is the fact that, within the past few years, we earthlings have developed for the first time the capacity to receive coherent signals from outer space. Moreover, our capabilities are improving almost daily.) So let us assume that we have established contact with a civilization five light years away. This would mean that we could complete a two-way communication every ten

years. Perhaps the first twenty years would be spent trying to overcome the language barrier, with the language of mathematics providing the breakthrough. After that we would want to begin sharing personal information about each other. What do we look like? What is our means of reproduction? If sexual, how do the sexes differ in overall appearance and in all specifics such as height and weight? At this point the magnificance of statistical analysis would emerge. All that would be required to describe most human characteristics would be two bits of information on each: the mean and the standard deviation. Since the civilization living five light years away would almost surely have independently discovered the mean and standard deviation, imagine the wealth of statistical information we could convey by a series of short radio bursts.

Height
  Male: Mean =
    Standard deviation =
  Female: Mean =
    Standard deviation =

Practice of premarital interdigitation
  Male: Mean time =
    Standard deviation =
  Female: Mean time =
    Standard deviation =

Putting aside extraterrestrial communications, we can see that we humans could even communicate much information to each other if more of us understood the language of statistics. Let's look at a few examples that illustrate the use of the mean and the standard deviation when dealing with normally distributed scores. The three children of your Aunt Mathilda—Mary, David, and Laurie—have just taken standard educational tests in school. Their scores on each of the tests were as follows.

Spare Parts Identification Test: Mary's score = 80

Clerical Aptitude and Nomenclature: David's score = 70

Car Aptitude and Nomenclature Test: Laurie's score = 650

Aunt Mathilda knows that you are very smart so she singles you out for advice. "What shall I do with these scores?" she asks pleadingly.

Successfully fighting off the temptation to give her a smart reply, you answer instead, "Go to the *Mental Measurement Yearbook* and find the mean and standard deviation of each of these tests."

"But how will that help me?"

"Get me the information I requested and I shall explain," you whisper with a conspiratorial air. "The interpretation of test scores involves a closely guarded secret code that only a gifted few have managed to crack."

"Oh, how thrilling!"

Breathlessly, she returns a few hours later. Voice quivering with excitement, she reveals all: "The mean on S.P.I.T. is 100, with a standard deviation of 10; David's C.A.N. has a mean of 50 with a standard deviation of 10; and the mean of Laurie's C.A.N.T. is 500 with a standard deviation of 100."

"The first thing we must do now," you say, resuming your secretive air, "is to arrange each score with its accompanying mean so that we can make certain comparisons. So this is what we have."

Individual	Test	Score	Mean	Difference
Mary	S.P.I.T.	80	100	−20
David	C.A.N.	70	50	+20
Laurie	C.A.N.T.	650	500	+150

"Oh, huh."

"Now, if we subtract the mean from each score, we find out two things."

"What things?"

"For one, whether a score is above or below the mean. A minus score indicates below the mean. Second, we learn *how far* the score is above or below the mean."

"Oh, I see. Mary's difference score is −20 so she is 20 points below the mean. Laurie is 150 points above the mean. That means she did the best."

"Not necessarily. That's where the standard deviation comes in."

"Oh, that thing again. I don't have the foggiest notion what it is but it sounds kind of naughty."

"Naughty only if you don't understand it and it keeps you in the dark."

"Anything that keeps you in the dark can't be all bad."

"Actually, it's a lot like a car. You don't have to know how it's put together in order to use it. Now, listen closely. I am about to reveal the most intimate secret of the standard deviation. The first thing you do is divide each of those difference scores—also called deviations from the mean—by its corresponding standard deviation."

"Wait a minute, you've lost me."

"O.K. Step by step. Mary got a S.P.I.T. score of 80. Since the mean S.P.I.T. is 100, her deviation score is −20. Dividing this by the standard deviation of S.P.I.T. yields a final score of −2.00. This final score is called a *z*-score. Now David's C.A.N. . ."

"Rein up, cowboy, you're losing me."

"O.K. Let's go over the logic of transforming to *z*-scores. You have the scores of your children on three different tests. You know the means and standard deviations of each of these tests. Right?"

"Right."

"You can subtract the mean from each score to determine how far it is above or below the mean. Right?"

"Right on."

"Now you can divide each difference by its standard deviation. In doing so, you express the difference in standard deviation units."

"You just lost me."

"O.K., try this one. Wilt the Stilt is 84 inches tall. How tall is cousin Mary?"

"48 inches."

"How much taller than Mary is Wilt?"

"Well, let's see. That would be 84 minus 48 is 36."

"Good. Now, how do you express this difference in terms of feet?"

"There are 12 inches in a foot . . . divide by twelve. So he's three feet taller than Mary."

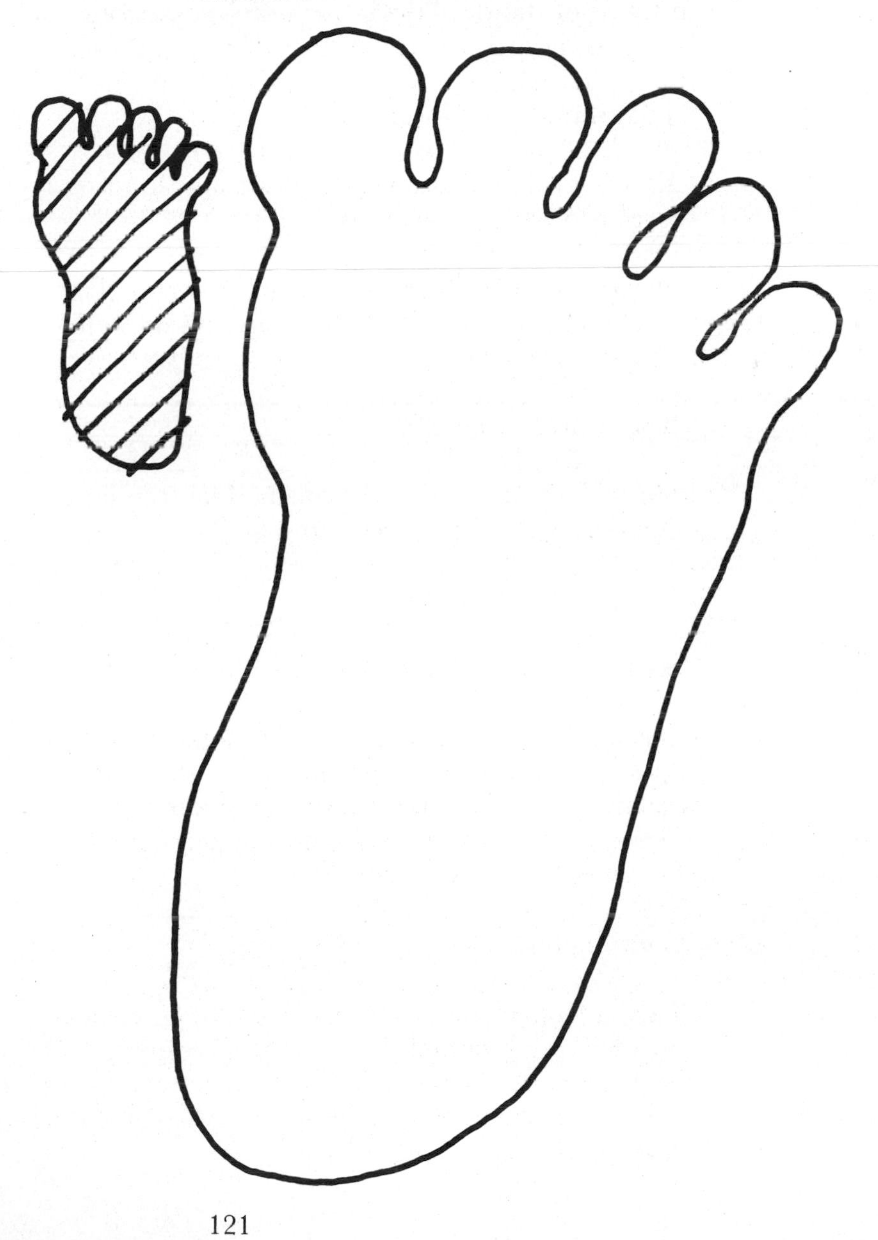

"Good, the difference between Wilt and Mary, expressed in feet, is three. Similarly, the difference between Mary's score and the mean, expressed in standard deviation units is -2.00. So you see, it's the same—to express deviations in inches in terms of feet, you divide by twelve. To express deviations from the mean in terms of standard deviation units, you divide by the standard deviation."

"That sounds simple enough."

"Now, let's do the same with David and Laurie's scores. Go ahead, you try it."

"David's score is 20 points above the mean. The standard deviation is 10. Therefore, his z-score is $20/10 = +2.00$."

"Beautiful. Laurie's?"

"She's 150 points above the mean. The standard deviation is 100. Her z-score is 1.50."

"Again, I say beautiful."

"So again I say, what do I do with it?"

Gnashing my teeth again to avoid a smart and expletive-deleted reply, I answered, "All you need now is a table that gives you the percentage of cases above a given score and the percentage of cases below a given score."

"Come again?"

"Take a look at this table. It's a simplified version but it will serve to make my point. In column A are the z-scores. Column B shows the percentage of cases that obtained lower z-scores. Mary got a z-score of -2.00, right?"

"Right."

"Looking at column B adjacent to the z-score of −2.00, we find that the percentage of cases with lower *z*-scores is 2. We call this percentage a percentile rank. Mary's percentile rank is 2, which means that she scored higher than 2 percent of the people taking this test. If you look at column C, you'll note that 98 percent scored higher than she."

"She didn't do so well, did she?"

"No, but maybe it's not of earthshaking importance that she does well in S.P.I.T. When glancing at her record a few months ago, I noticed that she scored 135 on a standard I.Q. test."

"That's good?"

"Well, you figure it out. The mean of this I.Q. test is 100 and the standard deviation is 16."

"Her deviation score is 135 minus 100. That's 35."

"Right."

"Now you divide by the standard deviation. Let's see. Her *z*-score is about +2.2. Hey, that means her percentile rank is 99.

"And that means?"

"Ninety-nine percent of the kids in her age group got scores lower than she."

"How about David and Laurie?"

"Well, David's *z*-score is +2.00. That means his

percentile rank is . . . let's see—98. And Laurie's is 93. Some kids I've got."

"Now on the three tests—S.P.I.T., C.A.N., and C.A.N.T.—who did the best?"

"Now wait a minute. You can't compare apples and oranges."

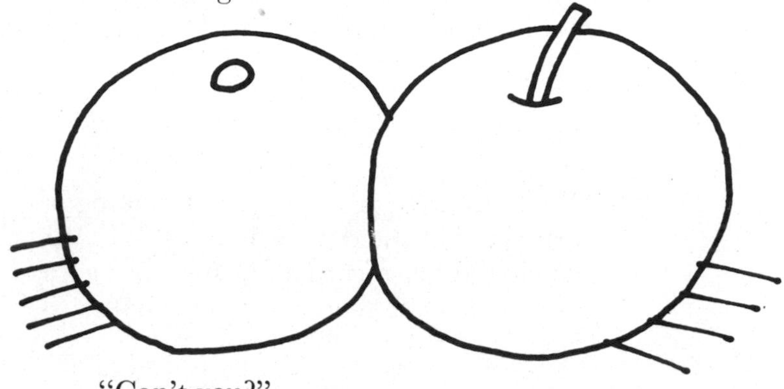

"Can't you?"

"Of course not. On second thought . . . well, of course. David's z-score on C.A.N. was the highest and Mary's S.P.I.T. z-score the lowest. In terms of relative performance on each test, David did the best, Laurie next, and Mary's S.P.I.T. was outrageous. You can compare apples and oranges. Yikes!"

"Well, Auntie dear, did you learn anything today?"

"You're damned right I did. Nobody's gonna snow me into believing test scores are magical numbers that can be interpreted only by the chosen ones."

"Anything else?"

"Statistics can be fun."

"You win the brass ring for that one, Auntie."

box 8.2 *So You Want to Interpret a Test Score?*

Here are the step-by-step procedures for taking all of the mystery out of the interpretation of test scores on standard psychological and educational tests.

1. Determine the mean and the standard deviation of the test. Sometimes different means and standard deviations are given for different age groups. Be sure to find these two measures for the age group in which you are interested. Sources of this information are the Administration Booklet for the particular test and the Buros Mental Measurement Yearbook. Since the Administration Booklets are not usually available to nonprofessionals, the Mental Measurement Yearbook is your best bet. If it is not found in your local library, it is almost certain to be in the collection of your nearest college or university library.

2. Transform the score you are interested in interpreting to a z-score using the following formula:

$$z = \frac{\text{score} - \text{mean}}{\text{standard deviation}}$$

If you are interested in interpreting a score of 40 and know that the mean and standard deviation are 30 and 9, respectively, you would have

$$z = \frac{40 - 30}{9} = \frac{10}{9} = 1.1.$$

3. Look up a positive value of 1.1 under column B of Table 8.1. Here we find an entry of 86. This means that 86 percent of a comparison group with which this score is being compared obtained scores lower than 86. Only 14 percent (column C) scored higher.

There it is. It's as easy as that.

Now a few words of warning. Don't treat the test score as fixed and unchangeable. We have a nasty habit of pigeonholing people on the basis of test scores. "We have found your score for ever and ever. Now we'll put you in your little box and let us hear no more from you." This is truly the tyranny of testing! In truth, a test score tells us something about your level of performance at the time you took the test. Hopefully, it reflects some of your long-term and prevailing characteristics. Otherwise, there is not much sense in submitting to testing. However, we must never lose sight of the fact that there are many transient forces, external as well as in-

table 8.1 Percent of Scores Above and Below a Given z

A z	B Percent of cases below	C Percent of cases above	A z	B Percent of cases below	C Percent of cases above
−2.2	1	99	0.1	54	46
−2.1	2	98	0.2	58	42
−2.0	2	98	0.3	62	38
−1.9	3	97	0.4	66	34
−1.8	4	96	0.5	69	31
−1.7	4	96	0.6	73	27
−1.6	5	95	0.7	76	24
−1.5	7	93	0.8	79	21
−1.4	8	92	0.9	82	18
−1.3	9	91	1.0	84	16
−1.2	12	88	1.1	86	14
−1.1	14	86	1.2	88	12
−1.0	16	84	1.3	91	9
−0.9	18	82	1.4	92	8
−0.8	21	79	1.5	93	7
−0.7	24	76	1.6	95	5
−0.6	27	73	1.7	96	4
−0.5	31	69	1.8	96	4
−0.4	34	66	1.9	97	3
−0.3	38	62	2.0	98	2
−0.2	42	58	2.1	98	2
−0.1	46	54	2.2	99	1
0.00	50	50			

ternal, that continuously assail us. Sometimes a score may be the offspring of these temporary affairs rather than the progeny of enduring internal relationships.

Another thing. Read very carefully the description of the standardization group. This is the group of individuals on whom all the test norms or standards are based. This is also the reference group against which all scores are compared. Most tests used in this country are standardized on middle-class, white individuals coming from English-speaking families. Also, the standardization groups for some tests are highly specialized, e.g., students graduating from four-year colleges. Obviously, the interpretation of a score must take the characteristic of the standardization group into account. A percentile rank of 50 on a mathematics aptitude test is very high if the standardization group consists of Ph.D.'s in mathematics!

Chapter Nine: We've Been Going Together for a Long Time but You Still Don't Turn Me On

I have a dear friend who is gifted with a marvelous sense of humor and perfect timing to go along with it. At one time she was rather pleasingly plump and about 5′4″ tall. She no longer is (plump, that is—she's still 5′4″ tall) since going on one of these fad diets. She likes to tell the story about one of her visits to the family doctor for her semiannual checkup.

"Get on the scale, dear. Let's see if you're being a good girl. Hm, 144 pounds. And how tall are you?"

"Six foot two, doctor."

"Six foot two? You're not anywhere near six foot two!"

"You want I should be overweight?"

This story can be used to illustrate many points. However, I have included it to explain something that we have all observed in everyday life—scores or values on two variables often go together. In the parlance of statistics, they are correlated. Tall people tend to be heavier than short people. Conversely, short people are usually lighter than tall people. And those in between on height are usually in between on weight. Adults high on intelligence are usually high on income, credit ratings, reading ability, and, for all I know, maybe even fat!

The fact that variables go together—or are correlated—provides the scientist with a powerful tool for predicting one event from knowledge of the other. If we know such things as barometric pressure, relative humidity, and/or wind direction, we can predict weather with an accuracy far greater than we usually admit when we roast TV weather persons over the coals. If we know a student's high school academic record we can make a pretty good guess about his or her performance in college. If we work in a mail-order house and we weigh the incoming mail each day, we can make a reasonably accurate prediction concerning the number of orders that will be contained therein. By making certain assumptions concerning the number of cars that will be on the road, AAA can arrive at a frighteningly accurate guesstimate concerning the number of traffic fatalities that will occur on a given holiday weekend.

The way in which two variables are related can be shown graphically, as in Fig. 9.1. This figure shows the expected range of weights for small-bodied women (as judged by glove size) at varying heights. Note that a band of weights is given for each height. This is because height and weight (like most other pairs of variables that scientists study) are not perfectly related. Some tall people are lightweights, while occasionally one finds a short person who is as wide as he is tall. Such aberrations are sufficiently rare that they are celebrated in song. "Mr. Five by Five" was a hit parade tune that I recall from a youth misspent in listening to swing (as my generation called it) or noise (as perceived by the

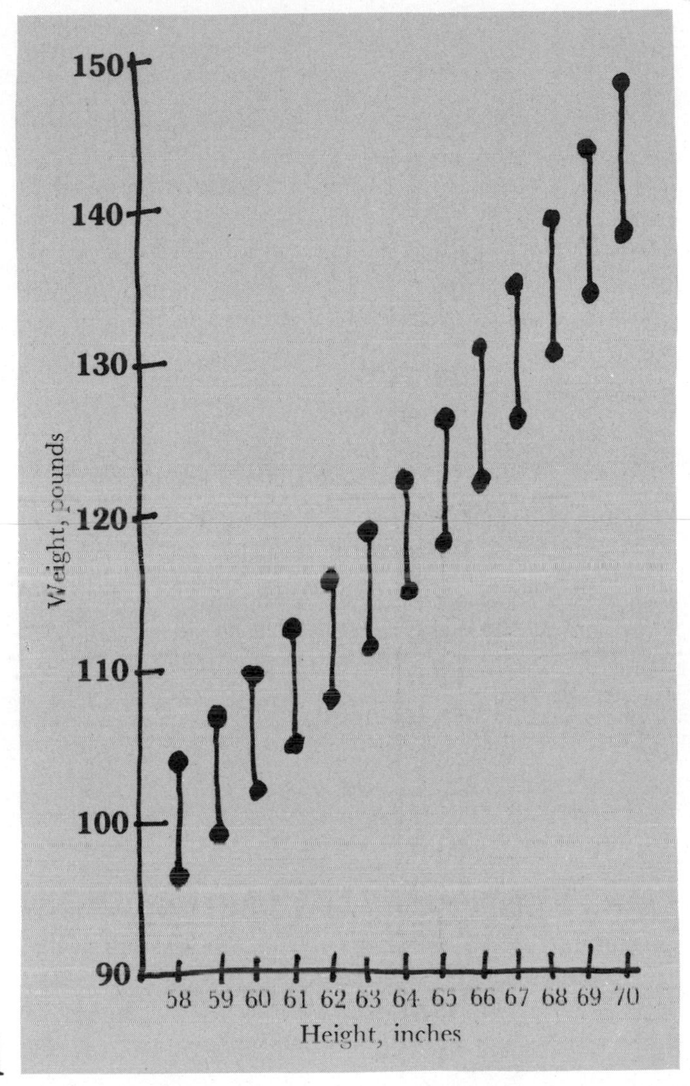

fig. 9.1

Graph showing the expected range of weights for small-bodied women at varying heights.

fig. 9.2

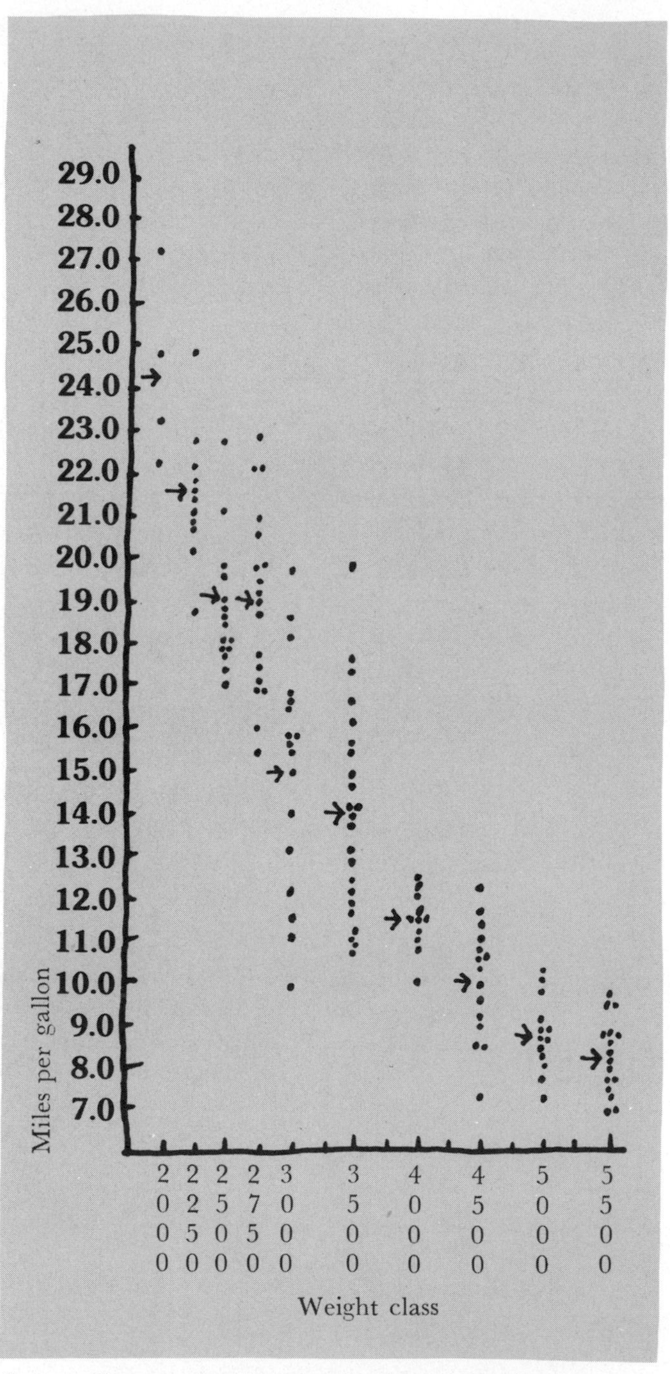

Scatter diagram showing the distribution of miles per gallon for automobiles of various body weights.

jaundiced ear of my parents' generation). Now if you really want to hear noise, listen to the stuff our kids are deafening themselves with nowadays. But I am wandering.

The point is that many events in the world about us vary together or, in the language of statistics, are correlated. The correlation may be direct (positive) or inverse (negative). The weight diagram given earlier (Fig. 9.1) is an example of a positive correlation. We saw that physical size and weight tend to go together. The relationship is called direct or positive because people at the low end of one scale (height) tend to be at the low end of the other (weight), those in the middle of one scale are generally in the middle of the other, and finally, those at the high end of one scale are often at the high end of the other.

In an inverse or negative relationship, the opposite is true. Take a look at Fig. 9.2. This is a scatter diagram of the relationship between the weight of an automobile and the number of miles per gallon obtained in simulated tests by the Environmental Protection Agency. The arrows indicate the mean miles per gallon for each weight class. This diagram makes vivid the extent to which the large car is a gasoline guzzler. Note how the big bullies of the road are feeble onlookers in the MPG derby. The lightest cars obtain, on the average, about 24 miles per gallon whereas the 2¾ ton monsters do well if they chug along for nine miles on each gallon of the liquid gold. On ten gallons of the stuff a small car can go from New York to Boston and still have enough left over to visit Folley Square. But the guy with the big car . . . forget it. He'll be digging in the wallet for a refill somewhere between New Haven and Hartford, Connecticut.

The beauty of correlated data, particularly when the relationship is high as in the case of weight

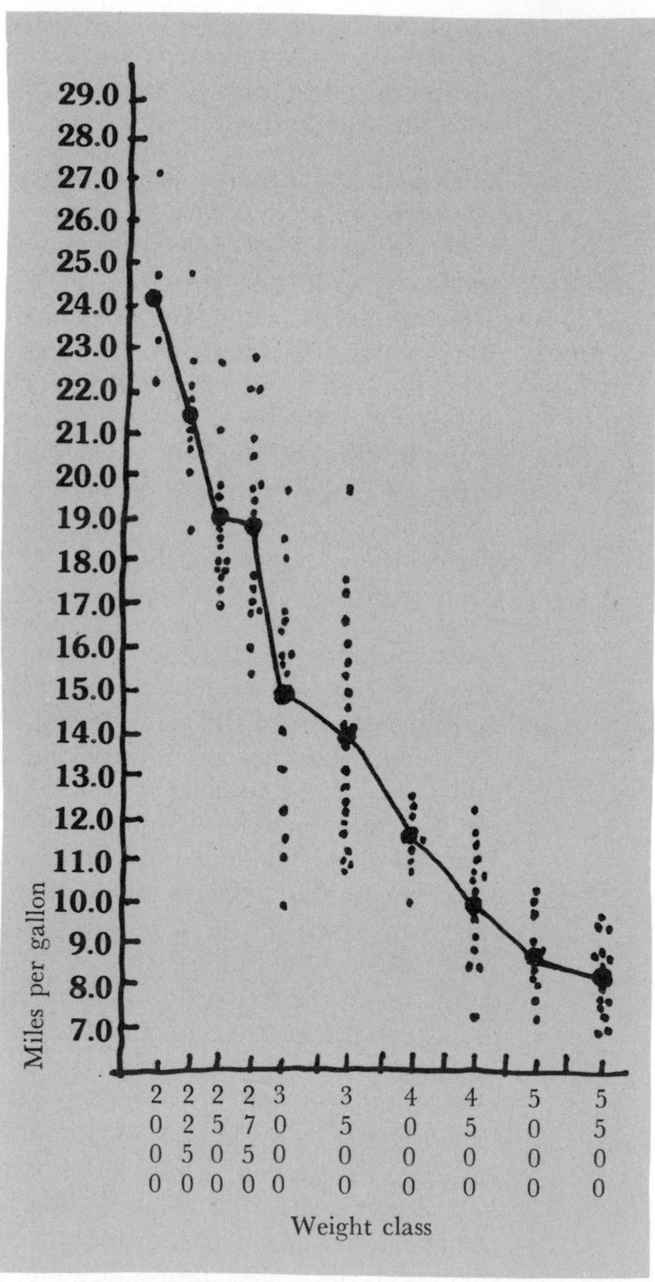

fig. 9.3

An approximation of the regression line for predicting miles per gallon from the weight class of automobile (regression of y on x). The line is obtained by connecting the points showing the mean miles per gallon for each weight class.

of auto and miles per gallon, is that mathematicians have worked out ways of predicting values on one variable from knowledge of the values of a correlated variable. The method is known as *regression analysis.* Please don't let the term throw you. We can arrive at a pretty fair comprehension of regression without stumbling about in the arcane caverns of mathematics.

Look again at Fig. 9.2. Now let's draw a line that connects the mean miles per gallon at each weight of car. The resulting line (Fig. 9.3) is a pretty good *approximation* to what mathematicians call the regression line for predicting Y-values (miles per gallon) from knowledge of X-values (auto weight). For purposes of discussion, we'll treat that line as if it were the regression line. In mathematical shorthand, it is called the line of regression of Y on X. But that is not important. What is important is that when the relationship between X and Y is high, we can use the regression line to predict Y-values from known X-values and achieve startling degrees of accuracy. (We could also predict in the opposite direction by finding another regression line, the line of regression of X on Y. But that's another story.)

Let's see how this works. Let us say that you are considering buying one of two cars. Brand A is flaming red and weighs 3200 pounds when equipped. The second, brand B, is a metallic gold and weighs 2400 pounds. You can make your best guess about the overall performance of brand A by looking at Fig. 9.3 and drawing a vertical line at 3200 pounds until it intersects the regression line. Now look left to find the corresponding value for miles per gallon. Your best guess, then, is that the 3200-pound car will average somewhere around 15 miles per gallon. Repeat the same procedure for brand B, drawing a vertical line at 2400 pounds. That comes to about 20 miles per gallon, an improved fuel performance of about 33 percent, using 15 mpg as the base.

So there it is. Regression analysis can be relevant to your daily living. Indeed, although you may not be aware of it, you have probably provided a numerical value in more than one regression equation in your lifetime, if you have ever applied for college admission, taken a series of tests for job placement, or filled out a life insurance application.

At the same time that regression analysis can be a tool of enormous value in the hands of a knowledgeable person, it can be an instrument of tyranny in the hands of an unthinking clod. This is what happens. The clod looks at the regression line and treats the predicted score as if it is God-given and absolutely precise. He says such things as, "Johnny Soinwhich got a score of 560 on the SATs. Since our cut-off score for admission is 570, Johnny is obviously incapable of performing at the intellectual level demanded of our students. Now Mary Inlikeflynn is a different story. She got a score of 580. Good girl. She'll do well at Dolc University." Our friend the clod forgets that all the data points fall directly on the regression line only when the correlation is perfect. In the real world, correlations are almost never perfect. This means that the data points are scattered about the regression line. (Take another peek at Fig. 9.2 to confirm what I am saying.) The lower the correlation, the greater the scattering, and the poorer the match between the prediction and the actual facts. Indeed, in any relationship that is less than perfect, about half of the actual scores will be lower than predicted and half will be higher. Thus, given the opportunity, Johnny Soinwhiches will occasionally perform admirably in spite of predictions to the contrary. And also on occasion, Mary Inlikeflynns will wind up on their tushes as they take the academic ten-count. I remember well my freshman year in college. A delightful, bright, and attractive coed informed me, "I don't have to study. I have an I.Q. of 135." Prior to the second semester of her freshman year, the dean generously offered to assist her in gaining admission elsewhere, after a suitable period of rustication.

All of this is another way of saying that there is a big difference between presenting a regression line (as in Fig. 9.3), complete with the dispersion of the original data points, and abstracting the regression line from the data (see Fig. 9.4).

The former permits the reader to observe the scattering of scores about the regression line, while the latter conveys the impression of great precision.

Correlation and Causation: But this is not the only way in which we can be deceived by correlational data. Consider the following facts and the conclusions drawn from them.

Fact 1: There is a positive relationship between the number of years that you go to school and the amount of money you are able to earn over a lifetime.

Conclusion 1: Therefore, get all the education you can. If you are a school dropout, you are also destined to become an economic dropout. On the other hand, if you get a college degree, your economic success is assured.

Fact 2: There is a positive relationship between the number of handguns produced each year and the number of murders committed by the use of this instrument.

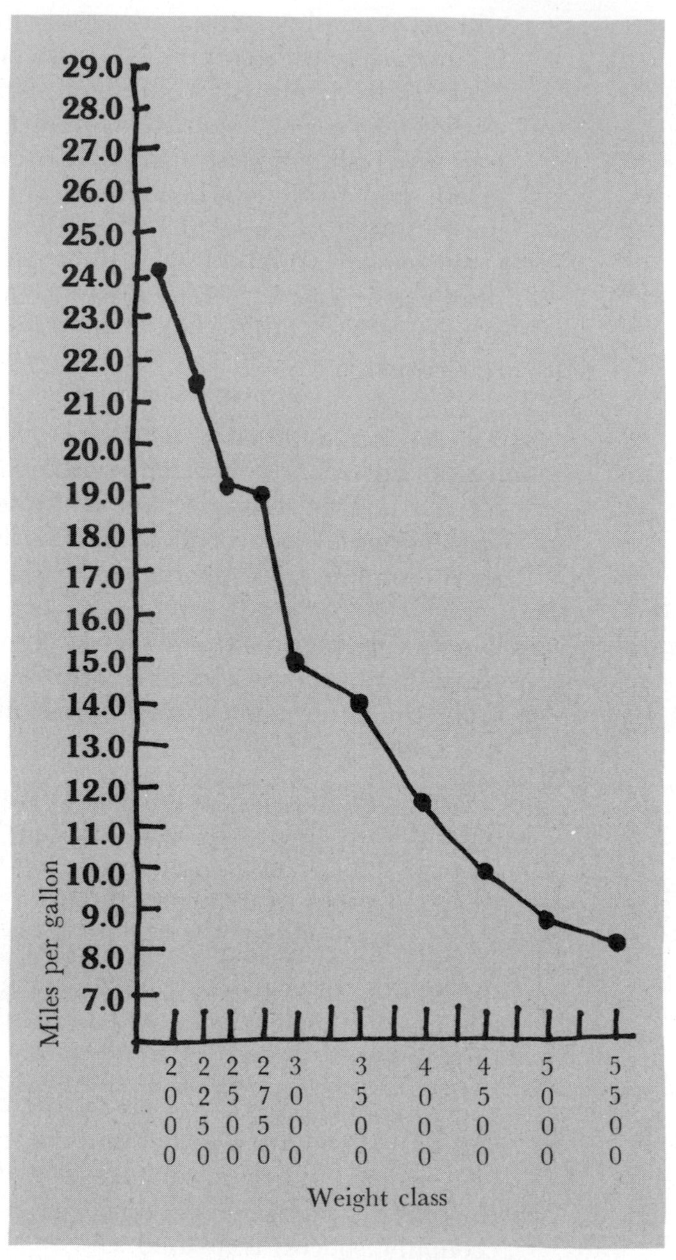

fig. 9.4

This figure is the same as Fig. 9.3. However, the data points have been omitted. The result is the appearance of greater precision than the data warrant.

Conclusion 2: We should reduce the annual production and importation of handguns since they are the cause of about two-thirds of all murders. Since fewer murders by handguns are committed when fewer weapons are available, we can lower the murder rate by restricting the production of handguns.

Both of these examples have something in common: They use the correlation between two variables as the basis for concluding that one causes the other. Let us examine each of them in more detail.

Figure 9.5 graphs the anticipated lifetime income for male wage earners, 25 years old and over, who have enjoyed varying numbers of years of education. The earning power of education appears most impressive. Men with four or more years of college can expect to earn, over a lifetime, about three times as much money as those who failed to complete elementary school. Surely there is no better evidence that education is a sound economic investment. Right?

Nonsense. Don't get me wrong. I have nothing against education. It has kept many pairs of shoes on the feet of my children over the years. Nor will I argue with the view that education provides many nonmonetary values that one may cherish often between matriculation and Medicare. In fact, I shall gladly assert the latter view. I still ruminate over outstanding lectures on philosophy, ethics, and psychology that I heard over 25 years ago during my undergraduate years. From my view, the value of education need not and should not be measured by an economic yardstick.

But many people do. Their arguments go like this, "Look, if you don't get a college degree, you'll always be a schlep like Joe Snopes. You wanna be a schlep all your life? Not able to provide for your family? Welfare? Or do you want to be like Stan Upright. He's got a good college education and he's going places, let me tell you that."

The only trouble is that the data do not prove that education leads to or causes higher incomes. The data merely prove that educated people earn more. They may be economically successful for many reasons: They may be brighter or more highly motivated; they may be consummate con artists; they may have well-to-do parents or grandparents (a Rockefeller is likely to earn a healthy income during the course of a lifetime if he possesses no other skill than the ability to clip coupons from municipal bonds.) Contrary to the view propagated in the Wizard of Oz, giving a college degree to a tin-head will not suddenly transform him into a computer capable of talking algebra. Stated succinctly, people who are educated differ in many ways from those who are not. Their economic success may stem from any one or a combination of all the ways in which they differ. It is no more appropriate to single out and credit education for the economic success of its graduates than it is to blame all of society's ills, as some do, on progressive trends in education or the "new math." (They learn that 4 + 2 is equivalent to 2 + 4 but do not know that either equals 6.)

But that's not all. Take another look at Fig. 9.5. Do you think that each of the columns is representative of the same population of wage earners with respect to all important variables? In this day of compulsory education, what age group is most likely to have failed to finish elementary school? Youngsters, middle-agers, or those breathing hard on FICA? Clearly, the last

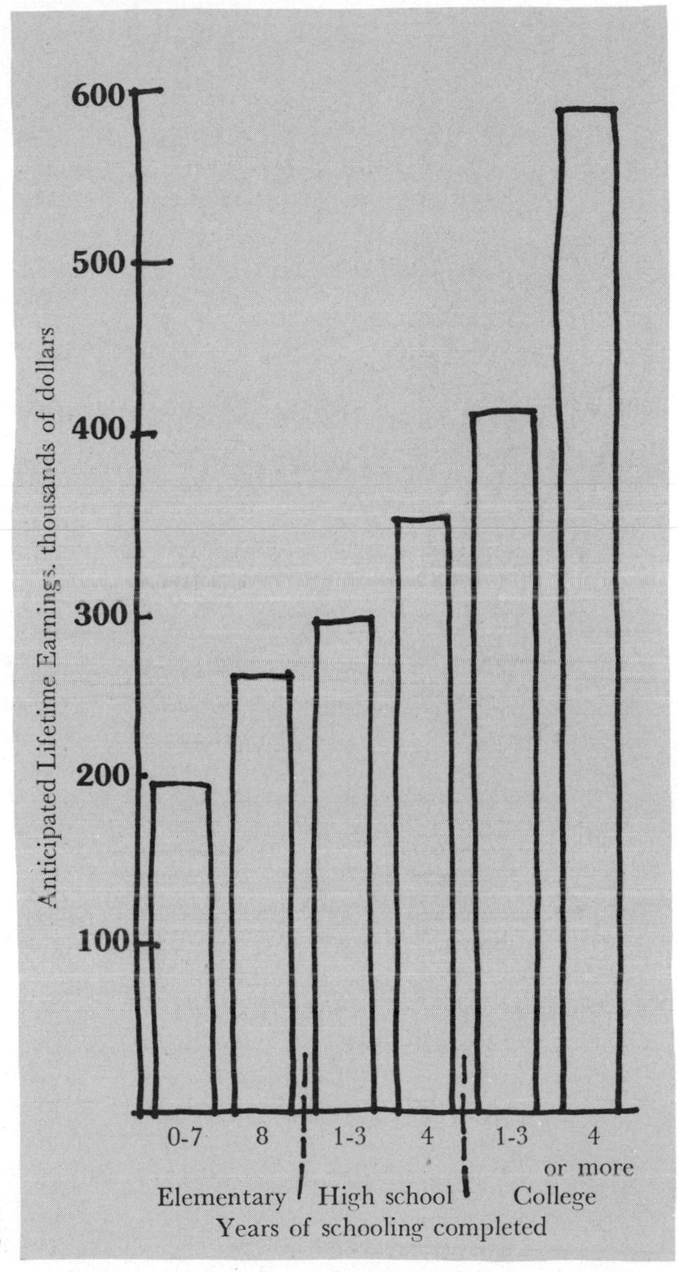

fig. 9.5

Expected lifetime earnings of male wage-earners 25 years of age and older. Over a lifetime, a college graduate can expect to earn about three times as much as a male who never completed elementary school. (*Source*: U.S. Bureau of the Census, Series P-60, No. 74.)

of the three. But they earned most of these wages prior to double-digit inflation, at a time when, as the saying goes, a dollar was worth a dollar. I even remember when twenty-five dollars a week was considered a living wage. Such citizens could not be expected to earn a whole lot of money over a lifetime, no matter how much education they had.

Now let's look at the righthand column of the graph. What proportion of our older citizens obtained college degrees? Very few, I assure you. So there you have it. Wage earners with low educational profiles are made up largely of older citizens who earned most of their income when wages were low. In contrast, those with high educational achievements are largely of the present generation, drawing income when wages are high. Their lifetime expectations are correspondingly high. So let's not justify education on pecuniary grounds. It simply doesn't wash.

In the above example, we are not justified in attributing a causal link between years of education and lifetime earnings because too many other variables are correlated with education—intelligence, motivation, and socioeconomic status of the family, to name a few. In the language of statistics, these variables are *con-founded*—they are so intertwined that there is no way of disentangling one from the other in order to ascertain their separate effects.

I am reminded of a story told to me by one of my friends in academia. It seems that one of his students had undertaken a term paper on the causes of alcoholism. After many weeks of intensive library study, the student uncovered a substantial difference among Catholics, Protestants, and Jews in their rates of alcohol addiction. It seems that Catholics are highest and Jews lowest. The student pondered these facts, asking such questions as, "In what ways do Jews differ from Catholics and Protestants?" Suddenly he was struck with the blinding light of cosmic insight, "Jews are circumcised; Catholics and Protestants are not. We can wipe out the scourge of alcoholism by circumcising everybody!"

To all of this, my professor friend raised but one question, "Do you think circumcision will be equally effective with female alcoholics?" In truth, Jews differ from Protestants and Catholics in many ways other than in the practice of the rites of circumcision—the ceremonial use of alcohol in early years, to name but one.

But what about the evidence, cited earlier, concerning the relationship between the annual production of handguns and the number of murders committed by gunfire? Surely, no one can question the validity of ascribing the increase in murder by guns to the greater availability of handguns. Or can one?

Let us examine the evidence. Table 9.1 shows the number and percentage of murders committed with various weapons between 1963 and 1973. There is no doubt about it. Aside from the automobile, the gun is our favorite weapon for abruptly terminating the lives of others. What is more, it is becoming increasingly popular with each passing year.

"But wait," you say. "Murder is on the increase, but so also is the population." Perhaps the two events are merely keeping pace with one another. So let's compare the rates of growth of both population and murder, using a technique we learned back in Chapter 6. We'll calculate fixed base index numbers between 1964 and 1973, using 1964 as our base year. The time index numbers are shown in Table 9.2.

When these index numbers are placed on a graph, the picture is quite clear. The rate of increase in murder is far outstripping the rate of increase in population. Although not shown here, it is also far outstripping the growth of population in that segment of the population that commits most of the murders (17-year-olds and older).

table 9.1 Murder victims, by weapons used: 1963–1973 (Adapted from *The American Almanac: The Statistical Abstract of the U.S.* New York: Grosset & Dunlap, 1976).

Year	Murder victims, total	% Guns	% Cuttings or stabbings
1963	7,549	56.0	23
1964	7,990	55.0	24
1965	8,773	57.2	23
1966	9,552	59.3	22
1967	11,114	63.0	20
1968	12,503	64.8	19
1969	13,575	65.4	19
1970	13,649	66.2	18
1971	16,183	66.2	19
1972	15,832	65.6	19
1973	17,123	65.7	17

So far we have established the fact that murder is on the rise and that guns are number one on the hit list. Now let's relate this increase to the number of firearms that are produced and imported each year. Presumably, greater production means greater availability and greater availability means greater opportunity and temptation. The graph is shown in Fig. 9.6. Pretty convincing evidence, huh?

'Fraid not. Please don't get me wrong. I am not trying to add ammunition to the arsenal of the "Have guns, will travel" crowd. Quite frankly, I think our insistence on the right to bear arms is a racial memory from the Neanderthal days. But my own feelings are irrelevant to the issue. If the weight of evidence is the high correlation between availability of guns and murder by gun, I must admit to being a skeptic. Why?

Just about everything—marriage, divorce, banana production, abortions—has been either

% Blunt objects	% Strangulations and beatings	% Drownings, arson, etc.	% All other
6	9	3	3
5	10	3	3
6	10	3	1
5	9	2	1
5	9	2	1
6	7	2	1
5	8	2	1
4	8	3	1
4	8	2	1
4	8	2	1
5	8	1	2

growing or decreasing since 1964. Any of these events will correlate with the increased availability of guns. Take a look at Fig. 9.7. It shows that the number of murders by *cutting* or *stabbing* increases as the number of available *firearms* increases. Does anyone seriously entertain the view that the greater availability of firearms increases the risk of death by stabbing or that we can reduce the number of gunshot murders by cutting down on the availability of knives? And how about Fig. 9.8? We see a beautiful negative relationship between the number of farm workers and the number of murders committed by guns. If we take this sort of evidence seriously, we had better figure out how we can get them back on the farm, even after they have seen Paree.

Errors of this sort are far more common than we would like to believe. Consider the following:

An aspirant for political office proclaims, "There has been a twelve-percent increase in crime

table 9.2 Fixed base index numbers between 1964 and 1973, using 1964 as base year. (*Source: The American Almanac: The Statistical Abstract of the U.S.* New York: Grosset & Dunlap, 1976).

Year	Murder by gun	Fixed base 1964	Population (in millions)	Time index fixed base 1964
1964	4,393	100	191.9	100
1965	5,015	114	194.3	101
1966	5,660	129	196.6	102
1967	6,998	159	198.7	104
1968	8,105	184	200.7	105
1969	8,876	202	202.7	106
1970	9,039	207	204.9	107
1971	10,712	244	207.1	108
1972	10,379	236	208.8	109
1973	11,249	256	210.4	110

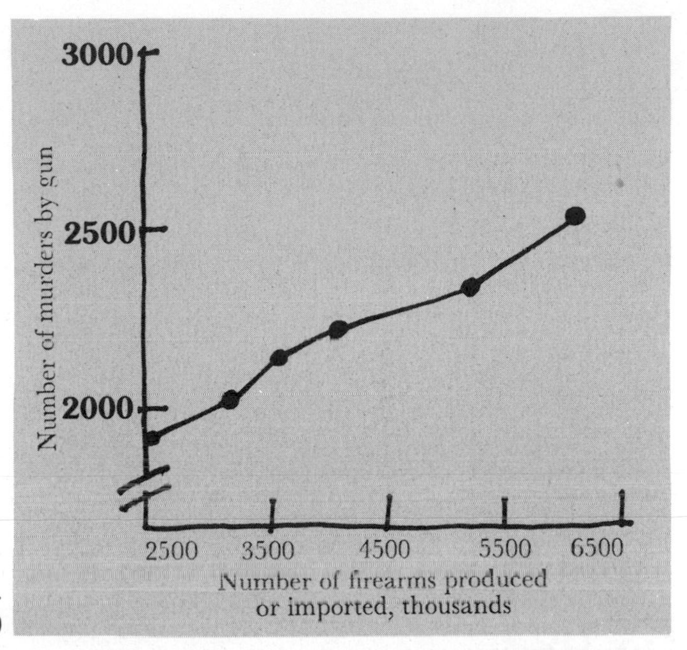

fig. 9.6

Relationship between number of firearms produced and number of murders by gun.

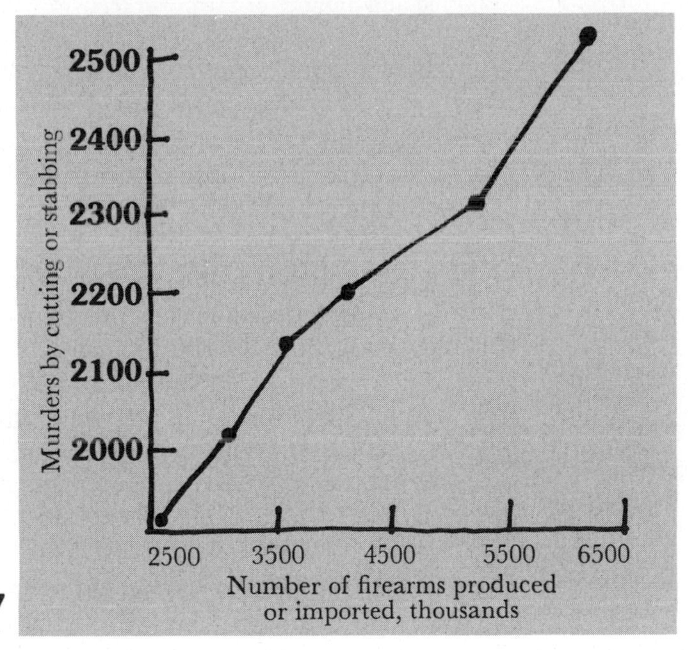

fig. 9.7

Relationship between number of firearms produced and number of murders by cutting or stabbing.

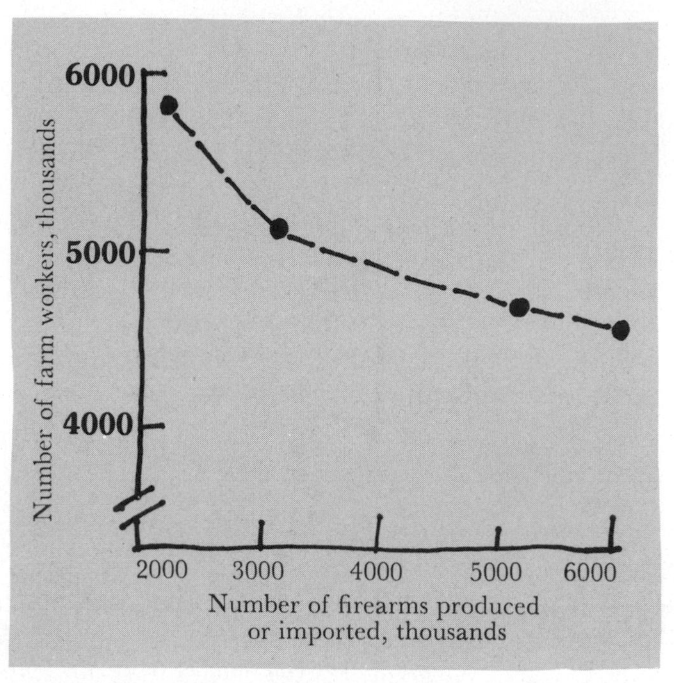

fig. 9.8

Relationship between number of firearms produced or imported and number of farm workers.

since my opponent, the incumbent, came to office." To ascribe a causal link between the incumbent and crime is to assert a degree of power that few individuals, political or otherwise, can rightfully claim.

"When you correlate the amount of advertising with the volume of sales, you find that large advertising budgets cause exuberant sales." However, it is quite possible that improved sales provide budget surpluses that permit greater advertising. It is not always clear what causes what.

---

box 9.1

Flush me a commercial

Social and behavioral scientists are constantly on the prowl for indirect measures of the incredibly broad range of human activities that they study. The reason is obvious. When dealing with topics on which the answers may reflect unfavorably on the the respondent's character, there is a pronounced tendency to conceal the truth with a tissue of exaggerations, half-truths, and outright lies. To the question, "How many joints do you smoke during a typical week?" a member of the counterculture may exaggerate his proclivities in order to demonstrate his disdain for law and order. In contrast, the married, respectable, and rising businessman, father of three children and a regular churchgoer, may respond with "a joint, what's a joint?"

If you were to conduct a house-to-house survey of television viewing, the educational channels would seem to dominate nighttime viewing. And how many homes do you imagine would admit to subscribing to such magazines as Playboy, Playgirl, Penthouse, *or* Oui? *Now if we could find indirect measures—measures correlated with what we want to assess but sufficiently different that the respondent's defenses are not aroused—we'd be able to get at the truth without offending anyone. One of my favorite examples of such an indirect measure is the Flushometer. What's the Flushometer?*

Some years ago a water district on Long Island puzzled over the fact that its demand for water during the evening hours was punctuated by short bursts of enormous activity followed by long periods of quiescence. At first the water company was at a complete loss as to what forces were orchestrating and synchronizing these visits to the family john. The picture that these nightly activities conjured up was eerie. Was some UFO out there in space exercising mind control and directing people to march robotlike to the throne? A possibility, however remote.

Then some bright-eyed individual got the idea that the toilet flushes might coincide with the commercial breaks on television programs. He followed up his hunch by recording the exact amount of water demanded during specific time periods in the evening and subsequently determining when the commercial breaks took place on the television programs being viewed during these hours. The results were most astounding. It was found that the greatest flushes occurred during the commercial breaks of those programs that ranked highest on the Nielsen ratings. In fact, over the first ten ratings, the correlation was

absolutely perfect. The number one program, which was I Love Lucy, *received the greatest surge of flush power during its commercial break and the next nine in order showed decreasing amounts of water consumed during their corresponding commercial breaks. In this one instance, at least, reasoning from correlation to causation would seem to be fairly straightforward and direct. Few would argue that the flushing of toilets was causing the commercials to go on, but one would probably be reasonably correct in hypothesizing that the commercials were sending people scurrying to the bathroom.*

Imagine the flush power if all TV programs synchronized their commercial breaks! End of the Energy Crisis.

"A number of species of birds appear to be endangered by persistent pesticides, such as DDT. Some eggshells are extremely soft and they are broken before the birds inside can hatch. The greater the amount of DDT in the yolk, the more likely it is that the shell is soft." One researcher found that both the amount of DDT found in the yolk and the softness of the shell correlated with the nearness to civilization. He proposed an alternative explanation for the soft shell phenomenon, namely, the birds close to civilization were more likely to be frightened by bird watchers (few of whom venture into remote locales) causing them (the birds, not the watchers) to lay their eggs prematurely.

There is a crippling form of emotional disorder known as schizophrenia. When blood samples from schizophrenic patients are compared with samples taken from normal people, numerous biochemical differences are observed. This relationship between biochemical events and schizophrenia has prompted some researchers to conclude that the disorder is caused by disordered biochemistry. However, an alternative possibility is that the onset of schizophrenia causes biochemical changes, perhaps due to the different life-styles of schizophrenics (e.g., different diets, different forms and degrees of socialization, different living conditions). To illustrate, some years ago a scientist in Scandinavia announced a simple test for the diagnosis of schizophrenia, one that could be administered as easily as the test for diabetes. The mental health field was aglow with suffuse excitement. Confirmation of his findings would have enormous implications since schizophrenia is generally acknowledged to be the most disabling of the emotional disorders. Alas, researchers in other countries were unable to confirm the results. Intensive investigation of the procedures revealed that a fairly standard method had been used in the selection of the control group. The control subjects, drawn from the "normal" population, were pretty well matched with the patients on such variables as age and sex. However, the patients in the hospital perforce took their meals in the institution; control subjects ate their meals at home. It turned out that the diet of the patients was deficient in citrus fruits. Suddenly it was realized that the scientist had, in fact, developed a diagnostic test. However, instead of schizophrenia, it detected vitamin C deficiency!

New York State recently banned total nudity, including bottomless entertainment, in establishments purveying alcoholic drinks. The reason? According to Liquor Authority spokesman, Michael Roth, "(because of) years of experience that have shown nude dancing and similar entertainment frequently lead to prostitution and other kinds of sexual conduct between performers and customers."* As an alternative hypothesis, what about the possibility that women willing to bare their bottoms in public might also be willing to share their bodies in private? It is also possible that the clientele of these establishments have other-than-visual entertainment in mind, and are somewhat more aggressive in pressing their point.

* Associated Press release, 5 December 1975.

The non sequitur that doesn't even follow

Way back in Chapter 2, I crowned David Janssen as King of the Non Sequiturs. Recall that, after citing a study proving Excedrin to be more effective for the relief of pain other than headache, *he joyfully concludes, "So the next time you get a headache try Excedrin." A non sequitur with as much subtlety as the noise level in the Eighth Avenue subway.*

But a non sequitur is not necessarily a blunderbuss. Some can be so subtle that they fail to arouse that little mechanism in the brain that cries out, "Tilt!" Lite beer has assembled an impressive team of retired sports greats—including Mickey Mantle, Whitey Ford, and Roosevelt Grier—who try to persuade us to guzzle the stuff with such arguments as "Lite beer is one-third less filling because it has one-third less calories." Now, being somewhat on the corpulent side, I'll buy the argument that a good-tasting, low-calorie beer must be ranked among the great achievements of the century. But less filling? That's something else again. What does the number of calories have to do with filling the stomach? If "low calories" equals "less filling," then we should be able to drink enormous quantities of the zero-calorie stuff (water) without getting filled. How many of you have attempted to chugalug a few tons of H_2O recently?

As professional as these non sequiturs are, they cannot be matched in subtlety with those arising from correlational data. As an example, take the figures recently released by the Statistical Bureau of the European Community which relate per-capita strike days to increase in the national product per capita and the rate of annual inflation.

What we see in Table 9.3 on the following page are two rather impressive relationships: a negative one between per capita strike days and increase in the per capita GNP; a positive relationship between per capita strike days and the rate of annual inflation. The conclusion? A national Sunday magazine expresses no doubt: "Do strikes serve social progress? Strike statistics from Germany, France, England, and Italy during the years 1968–1973 prove they do not" *(emphasis, mine)* (Parade, *28 December, 1975, p. 4).*

I find two things wrong with this conclusion. First, the writer is comparing apples to apple carts. Note that the strike days per 100 workers is related to the increase in the GNP of the nation as a whole. On the assumption that few labor strikes are called to promote social progress, it would be more appropriate to determine if strikes served to advance the living standard of the strikers. Such data are not revealed in the table.

The second thing wrong is already familiar to us —the assumption that strikes have caused *poor GNP and higher inflation. Equally plausible is the possibility that strikes are but another manifestation of the dissatisfaction and discontent that pervades a society that is not coping satisfactorily with its ills. When nothing else seems to work, why not try a strike?*

table 9.3

	Strike days per 100 workers, 1969–1974	*Increase of the national product per capita in dollars, 1968–1973*	*Rate of inflation (annual average)*
West Germany	240	3339	5.2%
France	901	2347	7.4
England	3035	1221	8.9
Italy	5083	1089	8.0

There is a lesson in all of these examples. Causation is not easy to establish. Correlation is a necessary but not sufficient condition for establishing a causal relationship. If events A and B are not correlated, it is hardly possible that one causes the other. The experiment is widely hailed as the one unambiguous means of establishing a causal link. In an experiment, two or more similar groups are treated alike in all ways except with respect to the administration of the experimental condition. If differences in behavior emerge, it is presumed that the experimental treatment caused the differences. To qualify as an experiment, one vital condition must be achieved: It must be possible to manipulate the experimental conditions so that they can be administered to any subject at random.

Unfortunately, many so-called experiments fail to qualify with this absolutely essential condition. The reason is that the so-called experimental treatment is often not under the control of the experimenter. It is not free to vary. In a sense, the subjects select themselves for the so-called experimental treatments. For example, in studies of intelligence of whites and blacks, it is not possible for the experimenter to say, "For the purposes of the study, I will make you a black and you a white. We will then compare your intelligence scores." In the real world, whites and blacks already exist as different individuals. They differ in so many ways that it is not possible to say whether or not differences in scores on I.Q. tests reflect differences in underlying intelligence or differences in prenatal care, diet, cultural values, motivations, language, or what-have-you.

Whenever you see the results of a so-called experiment, get in the habit of raising the questions, "Is it an experiment as defined above? Or is it, rather, a correlational study that masks under the guise of an experiment?" If you do, you'll find that an astonishingly large quantity of "conclusive results" are still very much in the speculative stage of their development.

Chapter Ten: Having Fun with Statistics -- The Probability Game

In Chapter 1 we took a brief, somewhat fractured look at the life history of statistics. Recall that the field took its first halting step in the gaming houses and shipping yards of European commerce. We also pointed out the two broad functions of statistical analysis: descriptive and inferential. So far we have looked at a number of descriptive measures—proportions, medians, means, standard deviations, and correlation.

The work of Pascal and Fermat was concerned with the so-called laws of chance, otherwise known as *probability theory*. It so happens that probability theory is the very stuff out of which inferential statistics is made. In a sense, then, when Pascal and Fermat were rescuing Chevalier de Méré from his fiscal embarrassment, they were also laying down the foundation for contemporary inferential statistics. To the uninitiated, probability theory may sound rather esoteric and foreboding. This chapter is dedicated to the proposition that it can also be relevant and fun.

Fifty-Fifty Ain't Always Even Odds

Try this one on for size. A friend approaches you with the following proposition, and asks, "Do you have four coins in your pocket?"

"What denomination?"

"Makes no difference."

"How about four unclad Kennedy halves?" (I just know you carry them around in your pocket.)

"That will do fine. Now I'll tell you what I'm gonna do. I'll make you a deal you can't refuse."

"Oh?"

"If you toss them up into the air and they fall to earth, how many heads do you expect?"

"Two. Two tails, too, for that matter."

"Good, I'll give you 1.2 to 1 odds you don't get two heads and two tails."

Would you take him up on that bet? If you said, "Yes," don't be dismayed when I tell you it's a sucker bet. Most people would fall for it. The odds would appear to be fifty-fifty or 1 to 1. In truth, the odds are really about 1.67 to 1 against obtaining two heads. In the long run your friend would stand to clean up quite a bit if he could find a sufficient number of people to fall for his scam.

And he probably would. That's because many people are befuddled by probability problems. They resort to intuition, and intuition can frequently lead a body down the primrose path.

Consider the following and circle whether you agree or disagree with the strategy.

You flip a coin six times. You are offered 2 to 1 odds that you will not obtain an even split (three heads and three tails). You take the bet. Agree? Disagree?

You flip a coin four times and it comes up heads each time. Since obtaining five heads in a row is extremely unlikely, you offer small odds that the next flip will not be heads. Agree? Disagree?

You are in a room with about 29 other people. A tall, thin man with slicked-down black hair and a fu manchu mustache states that two people in the room, at the very least, have the same birthday (day and month but not year). He offers even odds that he is correct. You accept the bet. Agree? Disagree?

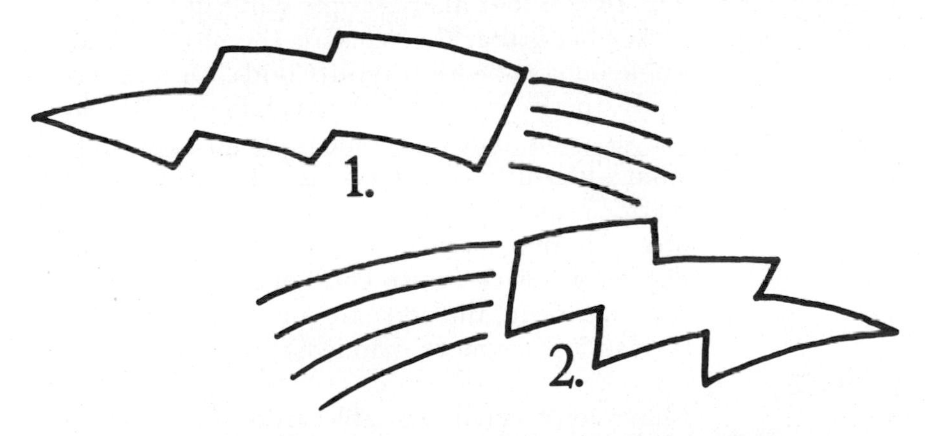

You are on a beach when an electrical storm breaks out. Noticing a tree that has recently been struck by lightning, you take shelter under it. "Lightning never strikes twice in the same place," you reason. Agree? Disagree?

If the probability of obtaining a six in a single roll of a die is 1/6, the probability of obtaining at least one six on three rolls is 3/6 or 1/2. Agree? Disagree?

Did you write disagree to all of these questions? Great if you did. But if you did not, please don't bury your head in shame. You are surely in good company with a whole bunch of fellow Americans.

Now let's look at each question in turn and see why you should have written "disagree" to each.

Will Six Coins Come Up Three Heads and Three Tails Fifty Percent of the Time?

The reason that many people will fall for this sucker bet is that they confuse the most probable single outcome with fifty-fifty odds. Or alternatively, they reason somewhat as follows: "If you toss six coins, half will come up heads, half tails. That's fifty-fifty or even odds." The truth of the matter is that the odds of an even split are fifty-fifty only under one condition—when you toss a single coin twice or two coins once. Surprising as it may seem, the odds are always against an even split when more than two tosses are involved.

I have constructed two tables from which you can obtain either the probability of a given event (Table 10.1) or the odds against the event (Table 10.2). Let's see how to use these tables for the present problem. Remember, we want to find out two things: (1) the probability of obtaining exactly three heads and three tails on one toss of six coins or six tosses of a single coin, and (2) the odds against this outcome.

box 10.1

Taking the edge off numbers

The values shown in Tables 10.1 and 10.2 are rounded to the third decimal place. What this means is that the values shown at the third decimal place are only approximate. However, the maximum error is only ± .0005 and in most cases, the error is less. Let's clarify this point with an example. In Table 10.2 we see that the probability of obtaining four heads in six tosses of a coin is .234. By alternately adding and subtracting .0005 to this number, we determine the "real" value of this number to be between .2335 and .2345. (In actuality, it is .2343750, so our real error was only .0003750.)

There are two types of situations in which we resort to rounding: (1) when we divide one whole number into a second and obtain either a long or infinite remainder, e.g., 1/3 = .333 333+; and (2) when we are dealing with variables that are never measured exactly, namely, continuous variables.

Let's take a closer look at the latter. When you measure such things as weight, height, or temperature, the measurements are only as accurate as the measuring instruments. What is more, it is always possible to imagine a more precise measuring instrument. For this reason every measurement is approximate. You may weigh 180 ± 0.5 on the bathroom scale, 180 ± 0.05 on the doctor's scale, or 180.05 ± 0.005 on a butcher's scale. Even with the most delicate balances in the world, weight would be rounded at some point.

Now what if you were to be told that the height of Mount Everest is 29,000 feet? What would you think? An estimate, rounded to the nearest 100 feet or so? Exactly. The People's Almanac

table 10.1

Probability of a given number of heads for varying numbers of tosses of a single coin ($n = 1$ through 10) or a single toss of varying numbers of coins ($n = 1$ through 10). *Example:* To find the probability of obtaining four heads on eight tosses of a single coin, find the column headed by 4. Read down to the row called $n = 8$. The probability is seen to be .273.

Probability of a given number of heads

	10	9	8	7	6
$n = 1$	—	—	—	—	—
$n = 2$	—	—	—	—	—
$n = 3$	—	—	—	—	—
$n = 4$	—	—	—	—	—
$n = 5$	—	—	—	—	—
$n = 6$	—	—	—	—	.016
$n = 7$	—	—	—	.008	.055
$n = 8$	—	—	.004	.031	.110
$n = 9$	—	.002	.018	.070	.164
$n = 10$	.001	.010	.044	.117	.205

table 10.2

Odds against various outcomes (numbers of heads) when a coin is tossed from one to ten times or when one to ten coins are tossed one time. Even odds is shown by 1:1. *Example:* You want to find the odds against obtaining five heads out of seven tosses (or five heads when seven coins

Odds against a given outcome (number of heads)

	10	9	8	7	6
$n = 1$	—	—	—	—	—
$n = 2$	—	—	—	—	—
$n = 3$	—	—	—	—	—
$n = 4$	—	—	—	—	—
$n = 5$	—	—	—	—	—
$n = 6$	—	—	—	—	61.5:1
$n = 7$	—	—	—	124:1	17.18:1
$n = 8$	—	—	249:1	31.26:1	8.09:1
$n = 9$	—	499:1	54.56:1	13.29:1	5.1:1
$n = 10$	999:1	99:1	21.73:1	7.55:1	3.88:1

5	4	3	2	1	0
—	—	—	—	.500	.500
—	—	—	.250	.500	.250
—	—	.125	.375	.375	.125
—	.062	.250	.375	.250	.062
.031	.156	.312	.312	.156	.031
.094	.234	.312	.234	.094	.016
.164	.273	.273	.164	.055	.008
.219	.273	.219	.110	.031	.004
.246	.246	.164	.070	.018	.002
.246	.205	.117	.044	.010	.001

are tossed at one time). Look down column 5 to where it intersects the row headed by $n = 7$. Here we find the odds against obtaining five heads is approximately 5.1:1. (All values in this table were derived from Table 10.1, in which the entries were rounded to the third decimal place. Consequently, the odds shown below are only approximate.)

5	4	3	2	1	0
—	—	—	—	1:1	1:1
—	—	—	2:1	1:1	2:1
—	—	7:1	1.67:1	1.67:1	7:1
—	15.13:1	3:1	1.67:1	3:1	15.13:1
31.26:1	5.41:1	2.21:1	2.21:1	5.41:1	31.26:1
9.64:1	3.27:1	2.21:1	3.27:1	9.64:1	61.5:1
5.1:1	2.66:1	2.66:1	5.1:1	17.18:1	124:1
3.57:1	2.66:1	3.57:1	8.09:1	31.26:1	249:1
3.07:1	3.07:1	5.1:1	13.29:1	54.56:1	499:1
3.07:1	3.88:1	7.55:1	21.73:1	99:1	999:1

(New York: Doubleday, 1975) tells an interesting story in this respect. On the back cover, it is claimed that Mount Everest is exactly 29,000 feet in height. However, reasoning that the public would consider this an estimate, the surveyors reported the height at 29,002 feet. A great story! The only trouble is that, in the body of this 2¾-inch thick volume, the height is given as 29,028 feet. If you can't believe an almanac, who can you believe?

The column to the the left of the table indicates the number of tosses. To find the probability of three heads on the face of six coins, read down this column until you find $n = 6$ (the number of tosses). Then read right until you find the column headed by 3 (the number of heads). At the intersection, you will find .312. This tells you that the probability is less than one-third that you'll obtain an even split (three heads in six tosses). In other words, an even split will occur by chance about thirty percent of the time. To find the odds against an even split, find the corresponding position in Table 10.2. Here we find the odds to be 2.21 to 1 against obtaining three heads and three tails. In the long run, the difference between 2.21 to 1 (the true odds) and 2 to 1 (the odds you have accepted if you wrote "agree") will drain you of finances. In fact, it is on the basis of such fragile margins that the gambling casinos in Las Vegas and Monte Carlo remain in clover. It is a basic fact of life that the probabilities are frighteningly dependable over the long run. When was the last time you heard of a casino going into Chapter 11?

---

box 10.2

In Chapter 1, I pointed out that the operation of the various devices in gambling casinos is among the most honest of human endeavors. This is not to say that the owners of such establishments are paragons of virtue. I don't know, nor do I care if they are or are not. It's just good business to run an honest establishment. Indeed, to run them any other way would be suicidal. So long as things are kept on the up and up, the management is guaranteed a certain and stable percentage of profit on the handle. But if something happens that changes the probabilities, the management suddenly becomes vulnerable!

Let us imagine that a die at the crap table develops a worn edge that makes certain events more likely to occur than others. (Incidentally, when this happens, the die can be said to have developed a memory. Events are no longer independent.) I am willing to bet that there would be some sharp cookie at the table who, noticing the unusual behavior of the die, would quickly appraise the new probabilities, and place bets that would have the house in fits.

However, this unhappy state of affairs is about as likely to occur as a January thaw in Prudhoe Bay. Gambling casinos are well aware of their vulnerability. As a result, consummate care and great expense go into selecting the casino's dice. To begin with they are beautifully balanced when put into service. Then, to protect the casino against the possibility of uneven wear, they periodically find their final resting place in the jaws of a crunching machine.

For the dice, at least, honesty doesn't pay. Their active lives last only a few hours, no matter how well they behave. But for the casino, honesty is a life and death matter.

Now that we know the probabilitics of thc various outcomes of the odds against these occurrences, how do we put them to our advantage?

Stated in other words, how do you calculate the *payoff function* of a wager?

Let's follow through with the same example that we have been using. You have persuaded your old friend (soon to become former friend) to accept even odds that three heads will appear on the toss of six coins. We have previously determined that the probability of this outcome is .312 and the odds against it are about 2.21 to 1.

To simplify calculations, we'll pretend that we have found a thousand suckers to take us up on the bet. We'll also assume that each risked one buck and wagered only once. On the average, our payoff will be as follows.

Step 1. In 1000 trials, we'll lose 312 times. We will thus pay out 312 dollars.

Step 2. Subtracting 312 from 1000, we find that we'll win 688 times. At a buck a win, we'll harvest 688 greenbacks.

Step 3. Subtracting our losses ($312) from our wins ($688), we obtain a net gain of $376.

Now depending on what base we use, we have a payoff of 37.6% (using the $1000 we risked as a base), 54.7% (using our total winnings as a base), or 120.7% (using the amount we actually paid out to cover the bettors' wins). I like the first of these better, since it relates payoff to risk. Also, it sounds more humane, particularly if you are wagering with a declining stable of friends.

But how do you make the calculations when you are offering odds other than even? Simple. Let's imagine that, to make the cheese more binding, you had offered your friends odds of 2 to 1. In other words, you would pay off two bucks each time your friend won.

Step 1. In a thousand trials, you will lose 312 times. At $2.00 a clip, your total losses will be $624.

Step 2. You'll still win 688 times, to rake in $688.

Step 3. Subtracting your losses ($624) from your wins ($688), you'll net 64 bucks.

If you had offered 1.5 : 1 odds, what would your payoff have been? (Answer: $220)

I must emphasize that these are only average figures—what you would discern as the trend after many repetitions of the "have-friend-will-rip-off" scam. Sometimes you would win more than $64 and sometimes less. In fact, I'll give you ten to one odds that you don't win exactly $64. Any takers?

---

box 10.3 *For bridge freaks only*

Even though Tables 10.1 and 10.2 describe the expected behavior of coins, their applications extend into the most important domains of human activity. The game of contract bridge is one example. Let us say that you have contracted for a certain number of tricks in a given suit. After the dummy is displayed, you have a complete count of the number of outstanding cards in each suit held by the opposition. All you lack is information concerning the way in which these cards are distributed between the two opponents. Table 10.1 permits you to calculate the approximate probability of various possible distributions. For example, imagine that you, playing South, have contracted for four spades. You have five trump in your hand and three more are in the dummy. What is the probability that you'll get a horrendous trump break and find all five outstanding trump in one hand? How does Table 10.1 help?

Go down the lefthand column for n = *5 (representing the five outstanding trump). Read across to the first entry. The value .031 tells you the probability of all five trump being in the West hand (and zero trump in the East). The next entry tells you the probability of four trump in the West hand and one in the East (*p = *.156); next three trump in West and two in East (*p = *.312); and so on until the last entry on the right (*p = *.031), which provides the probability of finding all outstanding trump in the East hand. To find the probability of a 5–0 distribution, merely add the probability of finding all five in the East: .031 + .031 = .062.*

The probability of finding

a 4–1 distribution is .156 + .156 = .312;

a 3–2 distribution is .312 + .312 = .624;

a 4–1 distribution or worse is
.031 + .156 + .156 + .031 = .374.

The resourceful bridge freak can find an almost limitless use of these tables. Among other things, they will assist in making decisions in the following cases.

1. In a suit headed by the AKJ, should I finesse or play for the drop of the Queen when there are (a) three, (b) four, (c) five, or (d) six outstanding cards of the same suit?

2. (a) With six outstanding trump, what is the probability of a 3–3 split in trump? (b) What is the probability of a split other than 3–3?

3. What if there are seven outstanding trump? Oh, I know. Good bridge players would never permit themselves to get caught in such a deadly contract. But with the quality of partners being what it is . . . Let's have the probability of

(a) a 4–3 split, (b) a 5–2 split, (c) a 6–1 split, or (d) a 7–0 split (perish the thought).
(e) What is the probability of a 6–1 split or worse?

4. Your contract depends on not finding West with four or more spades. With five outstanding, what is the probability that three or less will be in the West hand?

For the exciting answers to these thrilling questions, turn to the next page.

I'm Overdue for a Run of Luck (the Gambler's Fallacy)

Famous last words! This one takes many forms and invades many fields. The gambler is tossing coins and loses four in a row. Reasoning that five losses in a row is exceedingly rare ($p = .031$, odds against = 31.25:1; see Tables 10.1 and 10.2), he decides to increase the ante. "I'm due for a win," he proclaims confidently. Your favorite baseball player has gone 4 for 0. When he comes to bat for the fifth time, you exude confidence: "He's due for a hit." The opposing quarterback has just completed six consecutive passes against your team. You breathe a sigh of relief: "The next is bound to be incomplete or intercepted."

Although these examples are not all exactly the same, they have one thing in common—the belief that events have memories. It is as if the gambler is reasoning, "The coin will remember that it came up heads four times in a row and will try to balance out the 'law of averages' by coming up tails on the next toss." This is sheer nonsense. So long as the coin is "honest," it is just as likely to come up heads as tails on the next toss or on any toss, for that matter. We speak of this condition as *independence*—the outcome of one trial has no effect on later trials. This is just another way of denying that coins, or dice, or cards, or roulette wheels have memories.

It is somewhat different when dealing with activities involving behavior. Although a bat and ball have no memory, "turns at bat" are not always independent, particularly if they are against a pitcher of Tom Seaver's caliber. And your team could have seven passes completed against it because of its porous defense. Nevertheless, it is incorrect to cite the "law of averages" as the reason for expecting something different on the next trial. The law of averages has no enforcement agency behind it nor is there a Supreme Court to oversee the constitutionality of its "decisions." Indeed, the only thing it has in common with jurisprudence is its impartiality.

---

answer to box 10.3

Answers to: For bridge freaks only

1. (a) Play for the drop; $p = .67$ *that the Queen will fall (odds are 2:1 in your favor).*

(b) This gets rather complicated. If there is a 2–2 split, the Queen will fall. The probability of an even split is .375. However, it is also possible for the Queen to fall in a 3–1 split. Taking these possibilities into account, it turns out that both a finesse and a play for the drop are fifty-fifty propositions. However, play the Ace first on the off chance (1/8) that the Queen will be a singleton.

(c) and (d) The play for the drop is less than fifty-fifty. In the absence of other information, take the finesse. However, play the Ace first in the hopes that the Queen will be singleton.

Then take the finesse.

2. (a) .312; (b) .688

3. (a) .546; (b) .328; (c) .110; (d) .016; (e) .126

4. .811 or about 4 to 1 in your favor.

When Thirty Is Not a Crowd

Here's another example in which statistical intuition can lead a person far astray. Ask a person, "What do you think the chances are that we both have the same birthday?" He'll probably answer something to this effect: "Very slim, unless you know something I don't know." If more sophisticated, he'll answer, "About one in three hundred and sixty-five."

So far so good. But now ask him, "If we had thirty people in a room, what is the likelihood that two share the same birthday (day and month of year, but not necessarily the same year)?"

If he is like many people, he'll reason that the odds are thirty over three hundred and sixty-five or about one in twelve. It would thus seem as if the odds are almost eleven to one against this event. Would you believe that the actual odds are approximately two to one in favor of at least two people sharing the same birthday?

How can this be? We tend to forget that as we increase the number of people compared, we increase the likelihood of a match in a geometric rather than arithmetic progression. For example, if we have only two people, each can only match up with the other. The probability is one in three hundred and sixty-five. When we have fifteen people, each and every person can match up with fourteen others. With thirty people, each person has twenty-nine other people with whom to match birthdays.

In a teaching career that has lasted more than a score of years, I have frequently used this example in class. During the course of a semester, it is inevitable that the question of coincidence or a rare occurrence will arise during a class discussion. "My mother had a dream that something had happened to Aunt Mathilda. Sure enough, the next morning the phone rings. Something terrible had happened to Aunty M. the previous evening. She gained three pounds on her new low-calorie diet."

If I have at least thirty people in the class, I pull a bunch of pennies out of my pocket and offer to place bets—even odds—that at least two people have the same birthdate. I include myself since it increases the *n*. Many students will jump at the opportunity to beat me at something. For my part, I never had trouble finding enough takers to cover a cup of coffee.

The amazing thing is that I have never lost on this one, although I have done it more than twenty-five times. Of course, I covered myself by requiring more than thirty people in attendance before I would make my magnanimous offer. The point of this whole demonstration is that rare occurrences are not as rare as they sometimes seem, particularly if many people are dreaming bad things and a lot of bad things are happening in real life.

Incidentally, isn't it strange that, with 38 presidents of the United States, no two have had the same birthdate? Well, that's O.K. Gerald Ford was born on the same date—July 14 or the French Independence Day—as my younger son, Thomas. On the other hand, three of our first five presidents (John Adams, Thomas Jefferson, and James Monroe) died on our Independence Day—the Fourth of July.

Lightning Never Strikes Twice...

This one is really the gambler's fallacy with a twist. It takes many forms: Soldiers are advised to jump into a recently made shell-hole during a bombardment; during an electrical storm, you should seek shelter under a tree that has already been struck by lightning; when investing in stock, you should choose stocks that took a huge recent loss, because there is no way to go but up (Not so! Two of mine went into Chapter 11).

The twist is that if you follow this advice, you are probably *increasing* your risk. This is not to say that these events have memories. However, in a sense they can be said to have habits. Take lightning, as an example. It does not spray its bolts on the earth in a random fashion. Rather it seeks out a source to which it can run to ground. That source is usually a good conductor (such as a tree) that reaches above the ground (also a tree). For this reason, lightning often strikes objects on high ground. Thus, if an object is both on high ground and rises above the surrounding earth, it is a more-than-average candidate for a natural barbecue. For what it's worth, I can attest to this from personal experience. When I lived in the East, I was in an area reputed to be the second highest on Long Island (at a staggering 250 feet above sea level). The home itself stood about forty feet high and was perched on a knoll. Believe me, when it was struck by lightning during a violent summer storm, I did not breathe a sigh of relief afterward thinking, "Now that we got that one under our belt, we are safe." For good reason. A towering tree a few yards away from the house was struck by lightning the following summer. And I'd hate to tell you how many others landed within 1000 feet (about one second delay between flash and sound) of my home.

No, I would not recommend seeking shelter under a previously struck tree during an electrical storm. Nor, for that matter, could I be terribly excited about the prospect of entering water in which another person had recently been mutilated by a shark.

Do You Wanna Know Something? We're Both Impossible

Are you one of those people who always say, "I never win anything." Well, so am I. I have entered I don't know how many contests in a life time. I have never won yet. And it's not due to a lack of effort. I fill out every contest announcement that comes to my address, routinely refuse to buy whatever is being hawked, but, heeding the friendly advice, "You don't have to buy anything to be a winner," slap on a stamp and send it via the self-addressed envelope. But I never win (I'm overdue, though, you know. I'm really overdue. The next time I can't lose.).

I almost won once. While a small child during the depression, I used to scrape together pennies selling *Liberty Magazine* and the *Saturday Evening Post* to go to the Saturday matinee. Most of the time, I didn't go to see the movies—I shamefully admit I didn't care much for Greta Garbo or

Jean Harlow. I went to win one of the ten superlative gifts given to the holder of the lucky ticket. One Saturday, they called my number. They really did. I got all excited, screamed out, "I got it," and ran breathlessly to the stage. But, alas, in my great excitement, I had rubbed out the numbers on the ticket. That's the closest I have ever come to winning big.

But one thing we forget altogether too easily. We won the biggest gift of all—the gift of life. Have you ever calculated the odds against your number being called in that giant raffle? There were billions of male germ cells that joined in the race to fertilize the egg. Only one carried your very special chromosomal signature. All others would have been a brother or a sister, but not you. And how often does intercourse lead to the fertilization of an egg? What proportion of fertilized eggs survive? The odds against the particular combination that produced you or me are astronomical. But that is just the beginning. Each of the ancestors in our line of descendants was equally improbable. Thus, to produce either of us required the confluence of thousands upon thousands of extremely unlikely events stretched out over millions of years. If any one ancestor had altered his or her behavior in the very slightest—dated the redhead instead of Grandma, or complained of a headache—neither you nor I would be here today. If, at the beginning of time, a cosmic computer had been asked to calculate the probability that you or I would have been conceived on a specific date in the twentieth century, it would have replied, "The probability is so infinitestimal that we can safely say neither event will occur." In other words, we are impossible.

But then, so is everybody else.

And so is this book.

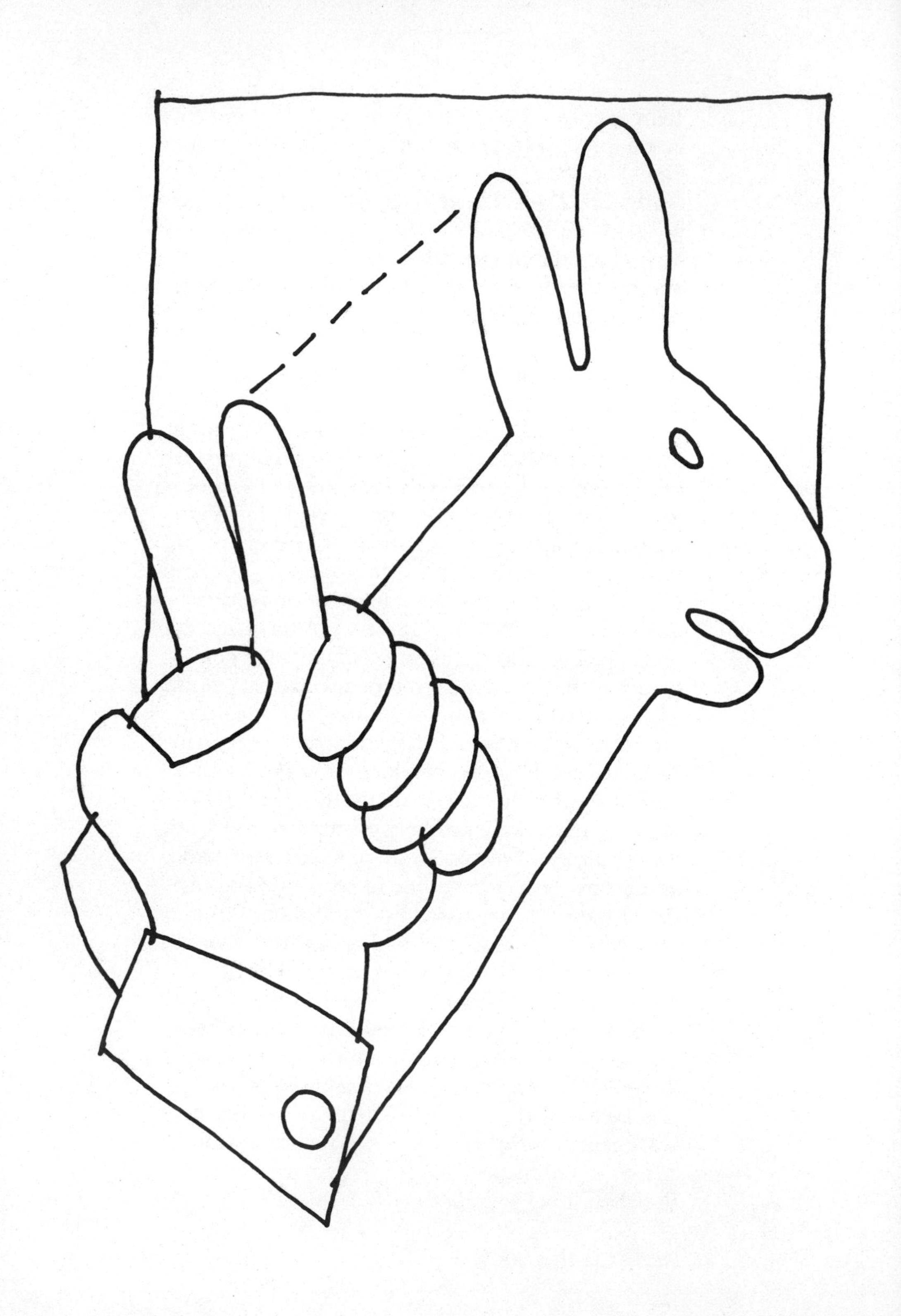

Chapter Eleven: Make Me an Inference

We have three dogs on our ranch on the outskirts of Tucson, Arizona. The baby of the three, Millie Muffin, is an eight-month-old black and white springer spaniel. Although mischief should be her middle name, she is a sheer delight. She is constantly exploring the desert flora and fauna with the indefatigable curiosity of a three-year-old human child. When spying something that moves on its own (such as a giant spider, better known as a tarantula), she leaps about two feet off the ground—ears and feathers flapping—and retreats several feet away, all the while barking like a fierce Doberman. But I know better. In reality, she is the world's greatest coward. I'll tell you how I know.

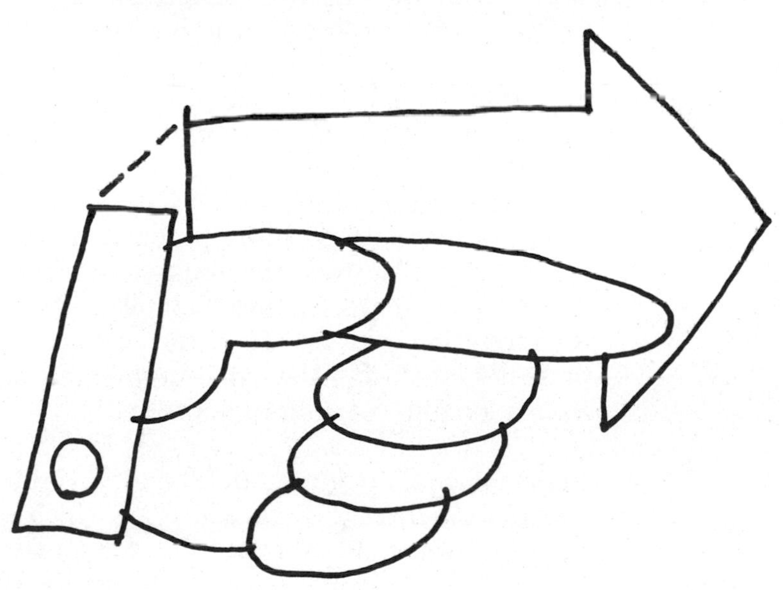

While she is prancing mindlessly about the giant saguaro, the ocotillo, the cholla, and other assorted exotic plants of the desert, I will sneak downwind of her. Then I will raise myself on the balls of my toes, stretch my arms in front of me, and shuffle my feet as I advance toward her in my best Frankenstein-monster style. When she spies me, she lets out a startled "yip," springs into the air, and begins a barking retreat. But all the while, her nose is probing the air, sniffing constantly, desperately drawing in samples in an effort to make inferences about the nature of the intruder. While observing Millie's antics, it occurred to me that the actions involved in taking samples and drawing inferences from these samples are among the most pervasive of mammalian activities. The pet dog, the family feline, the lion in the jungle, the gorilla in the rain forest, and the wife, husband, daughter, son, businessperson, doctor, lawyer, and Indian chief have this characteristic in common. They are continuously probing aspects of their environment, assessing the risks against the benefits, making probability judgments concerning alternative avenues of behavior, and pursuing those lines of activity that appear most likely to lead to desired goals. By this I do not mean to imply that all this sampling and probability assessment is conscious or deliberate. The truth of the matter is that nature has designed us all to be exquisite probability-generating machines. Without the ability to sample and thereby judge peril, or the availability of food, or the receptivity of a sexual partner, all species presently inhabiting the earth would have come from a long line of unborn ancestors.

In many ways, the inferential function of statistics merely formalizes the procedures by which we make judgments in everyday life. Let's take a look at a simple example and compare the reactions of a statistician with those of a just-plain-Joe or Josephine.

You have a friend, Al L. Hustler, who claims he has developed a secret process by which he can take a genuine 100-percent clad coin and convert it into a "heads" machine. "With this secret process, known only to me," he announces smugly, "I can make a coin come up heads over and over again. Just give me a quarter."

"Will a dime do?"

"Of course, but I'd prefer a quarter. It's larger, easier to read, and less likely to get lost."

"O.K. Here's a quarter. But don't forget where it came from." At this point you spy your friend, Phredrica Funk, by her own admission a world-renowned statistician. You explain Al's claim to her and ask her to evaluate the "experiment" he is about to launch.

"First," she says, "I want to examine the coin. Hm. 1976. A vintage year for clads. O.K. It looks on the up and up."

At this point, Al extracts from his pocket an object about the size of a matchbox. With a great show of pomp and ceremony, he places the coin head side up in the box. He presses a button on the side of the box and carefully watches the sweep second hand on his watch. "This must be very precise, you know. If I go a few seconds over, I'll bias the coin toward tails. I don't like making a jackass of myself."

"Jackasses are born, not made," Phredrica comments ascerbically.

Al ignores her obvious skepticism, seeming only to concentrate harder on the contents of his box.

"There, it's ready. Now I'll toss the quarter into the air . . ."

"Not so fast, Buster," Phredrica interrupts. "I'd like to see the coin you are about to toss."

"Don't trust me, huh?"

"Even presidents have been known to switch." She examines the coin intently. After what seems an eternity, she returns it to Al. "It's the same coin," she concedes.

"Now I throw it and catch it, and turn it over on the back of my hand. Presto, it's a head. Impressed?

Well, are you? "Not really," you reply. "You flip a coin once and it comes up either heads or tails. One or the other. It's a fifty-fifty proposition."

Phredrica, consulting Table 10.1 in this book, concurs. "Yes. Precisely a .500 probability of either a heads or a tails. May I please examine the coin again?"

Satisfied, she returns it to Al. He proceeds to toss it again and again a head raises its ugly. A smug grin passes across his face.

"I'm far from convinced," you observe. "But I must admit you haven't stubbed your toe yet."

"May I please see the coin again?" Phredrica examines it intently before returning it, stating, "Of course, what we have observed would occur twenty-five percent of the time by chance."

Another toss, another head.

"Well?" The look of imperial superiority on his face is infuriating.

"Three in a row is pretty good," you concede.

"Impressive but far from conclusive," Phredrica intones as she reaches again to take the coin from Al's hand. "Nevertheless, obtaining three out of three heads would occur by chance about 12½ times out of 100 trials involving three tosses of a coin." She utters these last words in a distracted fashion, as if her mind were many miles away.

"That's O.K.," Al confides. "I can keep this up all day if I have to."

Another toss, another head. "Well, what say you now?"

You have a great deal of difficulty containing your excitement. "Now I'm beginning to become a believer," you admit. Al's supercilious smile has somehow become less irritating.

Suddenly Phredrica's brain seems to snap back into place. She becomes all business: "Let me have another look at that coin." She scrutinizes the quarter, millimeter by millimeter, saying, "Now let's review what has happened. You put this coin in a box and said that you could do something to it that would cause it to come up heads on each toss."

"That's right, Professor."

"You have now tossed it four times and obtained four heads. Such an event would have occurred by chance only about six times in every hundred trials. That amounts to a probability of about six percent. I have examined the coin after each toss and found it to be the original. I know because I used my diamond ring to inflict a slight scratch on Washington's forehead."

"Clever of you." Al is now grinning from ear to ear.

She seems not to hear him, claiming, "Now, in science, all truth is probabilistic. We observe events or a series of events over a period of time. We work out ways of ascertaining how often, by chance, certain observations would be expected to occur. If we find that the probability is low that a given event occurred by chance, we are willing to rule out chance as an explanation. We then look for a lawful cause of what we have observed."

"How close am I to ruling out chance as an explanation?"

"Frighteningly. Some researchers are willing to rule out chance as an explanation if their observation would have occurred five percent of the time or less by chance. Others insist on a probability of one percent or less."

"What if I obtain another head on the next toss?"

"You will have broken the five-percent barrier. Technically, we refer to it as the .05 or 5% significance level."

"Well, here goes." He tosses the coin high in the air, catching it in midflight, and then reveals it for all to see. He smiles complacently. "Another head. Convinced?"

You nod assent.

While examining the coin, a distracted look has again come to Phredrica's face, "It's the same coin." Then, snapping her head as if to clear out the cobwebs, she continues, "No, I'm not yet convinced. You see, I'm one of the one-percent significance level people. You're down around the three-percent level now."

"And to reach the one-percent significance level?"

"If you get seven heads in a row, you will be below the one-percent level."

"Well, then, let's get to it." He tosses it again. Again a head. Phredrica examines the coin and returns it without speaking.

"Well, then, this is it. Let's give it a good ride." He flips it up by snapping his thumb. It arches high, turning lazily in the sun. Suddenly a dust devil spins to the site where we are standing. It catches the coin and moves it from its predicted course. Al circles like a baseball catcher trying to pull down a pop fly, lunges, and catches it just before it bounces on the turf. Sprawled on his abdomen, he displays the coin in the palm of his hand. "Well, what is it?"

"A head."

"That means I made the one-percent significance level?"

"Yes, you did, young man. I am impressed. I am quite willing to assert that the observations we made did not occur by chance."

"Well, there you go. We have just proved that my process works, right?"

Phredrica ignored the question. "Tell me, could you toss it just one more time?"

"Gladly. I'll do anything for science." Again the coin arches high in the air. But suddenly Phredrica is galvanized into action. Jostling Al aside, she reaches out and grabs the coin.

"Aha, just as I suspected," she expostulates exuberantly. "A two-headed coin!"

"A two-headed coin?" you call out in dismay. "But you examined it after each toss."

"No, I must correct you. I examined the coin he gave me after each toss. Our friend is obviously an advocate of the 'hand-is-quicker-than-the-eye' school."

For his part, Al seems undismayed, "You know, Professor, I have pulled that scam at least one hundred times. You're the first one who caught me. But it's all done in good clean fun. See ya." He smiles, but the cockiness is gone. He saunters off with an exaggerated gait.

You then turn toward Phredrica. "He almost pulled it off. You were about ready to concede that his box worked. Statistics can get you into a heap of trouble."

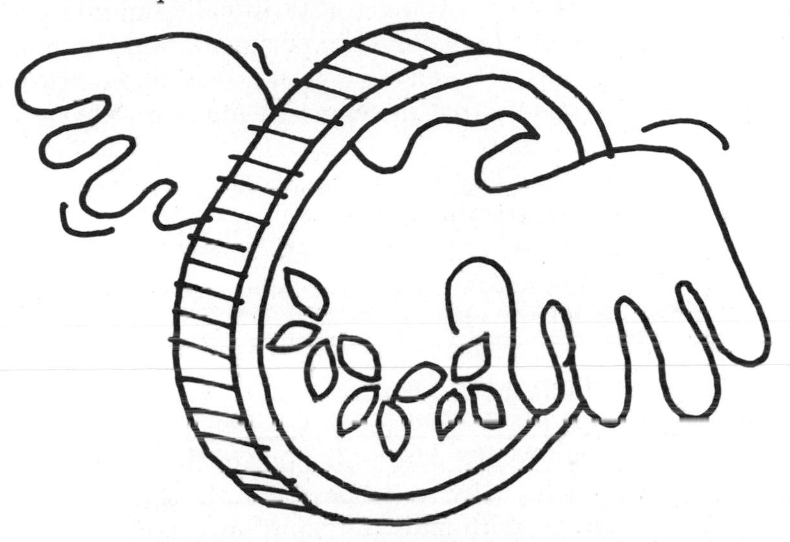

"You are right, but this is not one of those times. Inferential statistics only told me that our observations were unusual . . . not that the box worked. Now, several possibilities presented themselves. One is that your young friend had actually succeeded in building a device that can bias the outcome of a coin toss. But that doesn't exhaust the possibilities. Not by any means. A second possibility is that your friend was deceiving us in some way. In a sense, statistics was his undoing. Had I not been convinced by statistics that the coin was behaving in an unusual fashion, I never would have captured it in mid-air. This problem often comes up in research. Your statistical analysis convinces you to rule out chance as an explanation of your observations. You are willing to ascribe a cause to what happened but you are not always certain what cause to invoke. Most medical researchers agree that cigarettes are harmful to health. But are they to blame the smoke, the tars, the nicotine, the resins, the paper, or some combination of the lot? One of the values of a true experiment is that it limits the range of possible causes. Everything is kept as constant as possible except the experimental variable. If you get a statistically significant difference between an experimental and control group, you can state with a reasonable degree of certainty that the experimental variable *caused* the difference."

"And the third?"

"There is always the possibility that the event did occur by chance."

"But that's terrible."

"Not really. This is just one of the facts of scientific life. Just as the person in the street must learn to live with ambiguity and uncertainty, so also must the scientist. Every once in a while he or she will draw an incorrect conclusion and yet be blameless. The fact is that rare and unusual events do occur in this world of ours. When they happen during the course of an experiment, the researcher is almost sure to draw the wrong conclusion. But this is very true to life. Have you ever tried to run a traffic light that was changing from yellow to red? Most of the time, you won't get caught. The probability is low that a police officer will be at that particular street corner at the moment that you break the law. But once in a while the rare event does occur. That's what keeps traffic courts in business. And that is the same reason why scientists repeat so many of their studies before publishing the results. The big difference between the outside world and the realm of science is that scientists are well aware of this ambiguity and try to place a definite and known limit on their uncertainty. Probability theory provides those limits."

At this point, she looks at her watch, mumbles something about a meeting for which she was already late, and walks off at a brisk pace.

As she disappears in the distance, a disquieting feeling settles over you. You try to identify its cause. Is it the realization that science is fallible, subject to the same sorts of errors that all flesh is heir to? No, it's something more mundane than that. Suddenly you know. She pocketed your bicentennial quarter.

Chapter Twelve: You Can Prove Nothing Safe

Guess Which Hand?

Here's one to try on your friends. It is sure to mystify them, at least at the outset. Find two brand-new coins of the same denomination, and keep them always on your person. Then, when you are attending some social function that is beginning to lag, announce that you have found a means of thought control. Show the spectators one coin (keeping the other carefully concealed) and say that you will palm it in one of your hands. Their job is to guess the correct hand. Your job, of course, is to control their minds so that they will make an incorrect guess. Select one volunteer. Give him or her one of the coins to examine. This done, place it in the palm of one of your hands. Put your hands behind your back and assign one coin to each hand. Bring your fists forward and ask the volunteer to make a choice. Then answer, "I'm sorry, you're wrong," while displaying the coin in your other hand. (I once said, "You're wrong," and then proceeded to show the coin in the hand he had picked.) Repeat as often as necessary to send the volunteer scurrying for asylum from your thought control or until he or she catches on. I have experienced wide variation with this one. Children under six are particularly good victims. Some will persevere for hundreds of trials.

In any event, there is a lesson to be learned from this demonstration. It is exceedingly difficult to prove something by a process of elimination (in other words, if a coin is in the right hand, it cannot be in the left). Many people expect science to prove that something—a dye, a pesticide, a food preservative, a drug, pot, LSD—is not in the left hand (harmful) by showing that it is in the right hand (safe). In this chapter, I will show that this is a forlorn and hopeless expectation from either scientific or statistical analysis. The grim and inescapable fact is that it is relatively easy to demonstrate that something is harmful but impossible to prove that it is safe. Marijane will not someday be legalized because it has been proved safe. It will become accepted when and if the weight of evidence suggests that its harmful effects appear no greater than many other poisons we ingest or inhale daily.

But Dodoes Are Safe and Effective

Surely you've heard commercials on television that boldly proclaim: "Dodoes have been proven clinically safe and effective in the treatment of nagging back ailments," and quite possibly you accepted this statement without much question. The truth of the matter is, however, that nothing can be proved to be safe. You can prove things unsafe but the opposite—proving them safe—is not in the realm of possibility.

Now that's a pretty sweeping statement and I guess I had better be prepared to defend it. So let's get to it.

Let's start out by using a drug as an example, and see how we might evaluate its effectiveness. Let us imagine that the drug is supposed to be effective in the treatment of pains coming from the lower back. You identify a sample of people with

Dodoes have been proven clinically safe and effective in the treatment of nagging back ailments.

complaints of chronically aching backs. You randomly assign half the patients to one experimental condition and half to the other. You also use a double-blind study (see Chapter 4) in which neither you nor the patient knows what treatment is being administered. Of course, half the patients receive the drug and half the placebo (an inert ingredient that appears to be the same as the drug). Subsequent testing reveals that the individuals getting the drug actually experienced relief from pain in the lower back, whereas relatively few of the patients in the placebo or control group evidenced similar relief. Moreover, an analysis of the probabilities forced you to the conclusion that the greater improvement rate of the experimental subjects was not likely to be due to chance factors. To your satisfaction and that of fellow scientists, you have proved the drug effective. But that's the easy part. How do you go about proving it safe?

This problem, friends, is an entirely different ball of wax. Let's start out by asking the big questions: Safe for what? For whom? Within what time frame? At what dosage level? For what body system? How would you go about determining that something does not have any undesirable effects on any body system, either immediately after administration or at some unspecified future date? I'm waiting for your answer. Good. Basically what's done is that you look at the prior history of drugs of this sort and develop informed hunches about the locus of harmful effects, if there are any. You look for unfavorable effects in certain physical structures or certain enzyme systems or certain aspects of behavior. You must be selective, for reasons that will soon become obvious. You then set up a series of studies in which you follow your hunches. You look at the system or systems that you feel are most vulnerable to unfavorable reactions, if indeed there are any. You set up a double-blind experiment in which the drug is administered to the subjects at varying dosages and then you monitor the bodily or behavioral system that you have selected for study. If you find that the difference in the number of undesirable side effects among the experimental and control subjects is no greater than would reasonably be expected by chance, you feel confident that the drug produces no untoward effects.

So far so good. But have you proved that the drug is safe? Not by a long shot. Typically, failure to find evidence of deleterious effects is taken as evidence that the drug is safe. However, you have selected only a relatively few out of uncounted and, perhaps, uncountable systems for study. As we have noted, you made an informed guess. Your guess was that if the drug has bad effects, these effects will manifest themselves in system A or system B or behavior C or something of this sort. Your conclusions must be restricted to only those systems that you have tested, the dosage levels that you used, and the time frame within which the effects were studied. So you haven't really proved that the drug is safe in any general sense of the word. You have merely established that individuals getting the drug did not appear to have any untoward side effects when compared to individuals getting the placebo.

But then another question begins to intrude into your thoughts. Is it possible that the drug has an adverse effect on only a small segment of the population and that this effect could not be detected in a sample of the size that you used in your study?

When you say that a drug is safe with respect to certain physical systems that you studied, what you really have said is that you haven't detected any harmful effects among the sample of subjects that you've chosen. But what if only one in every one thousand people suffers a serious side effect? This fact would be completely obscured in the original research study. There would be no way of knowing it. This rare type of harmful effect will come to light only after many thousands of people have actually taken the drug in private use. Witness the birth control pill. Ten years ago, safe; five years ago, safe for most women; today, caution advised and frequent physical examinations.

But let's go a step further. We've examined only a few out of possibly a million different systems. For example, there are several thousand enzymes operating in the body. These enzymes control all aspects of our biochemistry. They are the precious templates that assist chemicals to line up in such a way that reactions can proceed efficiently. What if the drug produced an effect in an enzyme system that you aren't studying? If the effects were subtle, it would probably go undetected. Why? You simply did not design measurement of that system into your study. Remember the drug thalidomide? Who would have thought that a tranquilizer would have adversely affected some fetuses at certain stages of their development? Now with thousands of enzyme systems that could be adversely affected by any given drug, it is obvious that no drug could ever reach the marketplace if it were required that we check out the safety of every known enzyme system. And what about systems that have not yet been discovered?

But our problem does not end with enzyme systems. There are countless other systems operating in the body. What about the various nervous systems? How do you demonstrate that a drug does not have any deleterious effects on neural activity? How do you prove that it doesn't have an effect that is selective for a specific part of the brain? The brain contains literally billions of brain cells and clusters of cell bodies whose functions are largely still unknown. And this is only the beginning.

What about various aspects of behavior? A drug may not affect motor coordination but maybe it has some effect on memory, or maybe it doesn't have an effect on memory in general, but on some very selective aspect of memory. Or maybe it has an effect on the ability to solve abstract mathematical problems. How could one possibly study all the behavioral dimensions that might be adversely affected?

I've only begun to mention various systems of the body that could be adversely influenced by the drugs. So anybody who says that something has been proven safe is telling a lie. It may be told in ignorance, but it's still not true. There is no way that anything—not even water—can be proven safe. All we can say is that, given the sample of subjects and the measures that we chose to study, we did not find any evidence of undesirable effect. That's as far as we're entitled to go. This is not saying that there is anything wrong with science. And this is not saying that there is anything wrong with medicine or with the pharmaceutical houses. I am simply stating that it would be an impossibility to study all conceivable reactions to any given drug, or food additive, or pesticide, or chemical agent of any kind.

If you don't object to beating a dead horse, let's go one step further. Let's make up a figure. Let's say that there are 100,000 different systems in the body that are sufficiently distinct to be independently affected by a chemical agent. Now one might say, "All right, if you're really very, very careful and you have several lifetimes to spend on research, you might conceivably study 100,000 different bodily systems to find out whether or not a given drug is safe or, more precisely, that it has not produced evidence of any undesirable effects."

Now you run into the next problem. If you study a given system in isolation, you may not find any evidence that that system is unfavorably affected by a given drug. However, there is a concept that's very important in the field of statistics. It's called *interaction.* A given variable operating *alone* may produce one effect but when functioning in the presence of a second variable, the effect may be quite different. In other words, although $A = 1$ and $B = 2$, $A + B$ may not equal three. Take a simple illustration. You put fertilizer on your outside plants during a period of severe drought. The effect of the fertilizer is to burn all of the microfine root structures. Your plants die. But the same fertilizer, administered when moisture is available, causes the plants to absorb the nutrients in the fertilizer—instead of dying, the plants thrive. In other words, the effects of fertilizer administered without water are quite different from those of fertilizer administered with water. And you can get very complex types of interactions. For example, the effect of variable A may be dependent on the presence of variable B, but the effect of both may be dependent on C, D, E, F, or G. And if you have 100,000 systems operating in the body, it is quite possible that these systems are interacting in billions of different ways.

Again I state quite categorically, there is no way of studying all these various interactions. Now let me illustrate interaction with another example. As we are all aware, there are many additives placed in food. The claim is made that a given additive is safe. How is it found to be safe? As stated before, some astute researcher tested a series of different hunches about systems that would be sensitive to the undesirable effects, if there were any. But typically each system is tested in isolation of every other system. But what if A alone is safe with respect to a given system, and B alone is safe with respect to a given system, but

A and B together are lethal? Well, we're all familiar with the effect of barbiturates and alcohol. You take a barbiturate as a sleeping pill and it puts you to sleep. It has its desired effect. You take alcohol alone and alcohol is also a depressant; it makes you sleepy and you go to sleep. You take alcohol and barbiturates together and the sleep is sort of permanent.

One more example. There are many additives placed in the things we drink. For example, you wouldn't believe the line-up of additives that are put in beer. Some cause the beer to maintain a head for a longer period of time, others make the color lighter or darker, others halt bacterial growth, others maintain the flavor, and still others keep the bubbly going until there's no more beer to bubble. There are different additives for just about every effect that you could ever want. Why someday, they'll remove the beer and you'll drink the additives and notice no difference!

One additive that was rather thoroughly tested is a chemical called cobalt-sulfate. It is used in some European beers to extend the life of the foam over a longer period of time. The only trouble is that the tests for toxicity were conducted in isolation of the very beverage, beer, to which it is added. It so happens that alcohol plus this cobalt-sulfate interact in such a way that it forms a deadly poison for some people. Not for many, but for some. Over the years this combination has been incriminated in the death of at least 50 beer drinkers. That is why the labeling of the ingredients of all domestic and imported alcoholic beverages will be mandatory beginning 1 January 1978.

Alas, I won't even be able to drink a little booze without putting on my bifocals.

Index

Index

a

agriculture, 6
American Cyanamid, 7
Analysis of Variance (ANOVA), 7
average, 97, 102, 107, 111. *See also* mean; median; mode.
  ambiguity of, 95
  definition of, 95

b

bar graph, 57
base, stealing by the, 85
base year, 80, 82
bell-shaped distributions, 102, 113. *See also* normal distribution; symmetrical distribution.
biased sample, 76
biases, 59
Buros, 125. *See also* *Mental Measurement Yearbook.*

c

causality
  experiment and, 186
  inferential statistics and, 185
causation
  and correlation, 137, 154
  experiment and, 155
central tendency, 108
  mean, 96
  measures of, 95
  median, 96
  mode, 96
chance, 191
  inferential statistics and, 185
  and probability, 182

chance variation, 38
charts
  cumulative chart, 67
  Oh Boy! Chart, 60
confounded variables, 143
consumer price index (CPI), 79, 83
contract bridge
  and odds, 167
  and probability, 167
correlation, 129
  and causation, 137, 154
  DDT and soft eggshells, 151
  education and earnings, 137, 139
  examples of correlated variables, 130
  firearms and stabbing, 146
  handguns and murders, 137, 138, 143
  murder and population, 144
  negative (inverse), 133
  perfect relationships, 136
  positive (direct), 133
  prediction, 135
  regression, 135
  scatter diagram, 130
  scattering of data points, 136
  schizophrenia and biochemistry, 151
crime
  correlation between handguns and murders, 137
  farm workers and murder by gun, 146
  firearms and stabbing, 146
  guns and murder, 138, 143
  murder and population, 144
  rates, 81
  time ratios, 81
  violent crime rates, 82

cumulative chart, 67, 68
drawing original time/performance measures, 71
interpretive errors, 69

d

datum, 19
DDT, and soft eggshells, 151
demand characteristics, 43, 53
related to media, 51
descriptive (function), 14, 18, 157
descriptive statistics, headcount technique, 73
deviations
from the mean, 120
standard, 105, 115, 117, 118, 119, 120, 122, 125
difference score, 119
expressed in standard deviation units, 122
discrimination, by sex, 88
disembodied statistic, 25, 32
dispersion, 115. *See also* variability.
distribution(s)
bell-shaped, 102
normal, 115
skewed, 98, 102, 111
symmetrical, 102, 111
distribution ratio, 76
double blind, 47, 48, 53, 191
experiment, 192
Double-Whammy Graph, 65, 67
Douglas, Mike, 31
drugs
barbiturates, 197
and behavior, 194
and enzyme systems, 194
neural activity, 194
and normal activity, 194

e

educational tests
  interpretation, 125
  standard, 112
Energy Crisis, 29, 32, 109
Environmental Protection Agency, 133
  and miles per gallon ratings, 133
enzyme systems, 194
experiment
  and causation, 155, 186
  and demand characteristics, 43, 53
  double blind, 47, 48, 53, 191, 192
  and random administration of treatment, 155
experimental variables
  effects of, 38
  inferential statistics, 38
extraterrestrial communications
  and mathematics, 117
  and statistics, 117
extreme scores, effect of, 97

f

farm workers, and murder by gun, 146
Fermat, Pierre de, 3, 4, 157
Fisher, Sir Ronald A., 7
Flushometer, 149
food additives
  and health, 196, 197
  interaction and, 196
  and toxicity, 197
Ford, Gerald, 172

g

gambler's fallacy, 169, 173
gambling
  and odds, 165

and payoff function, 166
and probability, 165
Gosset, William, 6
graph
bar, 57
cumulative, 69
Double-Whammy, 65, 67, 79
pictograph, 62
rubber band boundaries, 57, 60
scatter diagram, 130
three-quarter-high rule, 59
time/performance, 67
graphic representation, purposes, 65
gross national product (GNP), 79, 153
and inflation, 154
and strikes, 154
Guinness Book of World Records, 114

h

Halley Life Tables, 5
head-count technique, descriptive statistics used with, 73

i

independence, 170
index numbers, assessing change, 79
indexes
consumer price index, 79, 83
cost of living, 66
gross national product, 79
purchasing power of the dollar, 66
inductive statistics, 20
inferences, and sampling, 178
inferential statistics, 15, 20, 38, 157
and causality, 185
and chance, 185

insurance, 5
interaction
  and food additives, 196
  of variables, 196
intuition
  probability and, 158
  statistical, 171

j

Janssen, David, 152
job discrimination, 88

l

law of averages, 169, 170
Life Table, 5

m

McCarthy, Clem, 20
mathematics, 117
  extraterrestrial communications, 117
mean, 19, 93, 96, 100, 103, 105, 117, 125
  calculated, 96
  deviations from, 120
measurement, 16
median, 96, 100, 110
  calculations of effect of extreme scores, 97
Mental Measurement Yearbook, 118, 125
Meré, Chevalier de, 3, 4, 5, 157
mode, 96
Morgan, Joe, 85

n

Nader, Ralph, 32
neural activity, 194
normal distribution, 115, 124

o

odds, 3, 164
  contract bridge and, 167
  gambling and, 165
  and life, 175
  table of, 163

opinion poll, 16
Orne, Martin, 42, 43, 44, 45, 46, 47
OY statement, 106

P

Paradoxical Percentages, 88
Pascal, Blaise, 3, 4, 5, 157
payoff function, 166, 167
Peak Shaving and variability, 108, 110
percent increase, 82
percent point increase, 82
percentage, 15, 74
  Paradoxical Percentages, 88
percentage of gain, 78
  ratios, 76
percentile rank, 123, 124, 125
  z-score and, 123
perception, 41, 42
pictogram, 62
pictograph, 62
placebo, 47, 191
  effects, 48, 49
poll, 12
  assessing change, 75
  head-count or survey, 74
  political uses, 75
population, 18
  and fixed base time indexes, 144
  murder and, 144
prediction
  and correlation, 135
  and regression, 135
probability, 3, 5, 164
  and chance, 182
  contract bridge and, 167
  gambling and, 165
  and intuition, 158
  samples and, 178
  table of, 162

probability theory, 157
  and science, 186
proportion, 18, 19, 74, 110
psychological tests, interpretation, 125
purchasing power of the dollar, 79
  indexes, 66

q

qualitative variables, 73
quality control, 6
quantitative variables, 74

r

random assignment, 191
random sampling, 18
ratios
  fixed base time, 80
  percentage of gain, 76
  time, 79
regression
  analysis, 135
  and prediction, 135
regression analysis, 136
  examples of misinterpretation, 136
regression line, 135, 136
Rocks, Lawrence, 29, 30
role
  of experimenter, 46
  of subject, 46
Roth, Michael, 152
rubber band boundaries, 57, 60
Runyon, Richard P., 30

s

Salk, 48
  vaccine study, 47
sample, 16
  biased, 76
samples, 14, 178

sampling, 14
and inference, 178
and probability, 178
scatter diagram, 130
and negative correlation, 133
and positive correlation, 133
schizophrenia, and biochemistry, 151
Scott, George C., 39
significance level
0.05 level, 183
0.01 level, 183
skew, 100
skewed distributions, 98, 102, 111
sleep
and alcohol, 197
and barbiturates, 197
standard deviation, 105, 115, 117, 118, 119, 120, 122, 125
standardization group, and test scores, 127
statistic, 19
statistical inferences
proving something effective, 190, 191
proving something safe, 190
statistical intuition, 171
statistical significance, 183
statistics
extraterrestrial communications and, 117
small-sample, 6
Stiller, Jerry, 31
symmetrical distributions, 102, 111

t

thalidamide, 194
three-quarter-high rule, 59
time/performance graph, 67, 71

time ratio
assessing change, 79
crime, 81
fixed base, 80
toxicity, and additives, 197

V

validating pseudograph, 55
variability, Peak Shaving, 108, 110
variables
confounded, 143
examples of correlated, 130
interaction of, 196
qualitative, 73
quantitative, 74

W

Wells, H. G., 9
Wilcoxon, Frank, 7
WOW statement, 105

Y

Yemani, Sheik, 98

Z

z-scores, 122
comparing relative status on different variables, 124
interpretation, 122, 123
and percentile rank, 123
transforming to, 120, 125